# Corrosion and Protection of Marine Engineering Materials

This book provides a basic introduction to the application of conductive polymers in corrosion protection, especially for marine environment corrosion protection research.

Conventional anticorrosive coatings, which are based on heavy metals such as chromium, zinc, and copper, are toxic to the environment. There has been a need to find suitable replacement coatings that are environmentally friendly as well as effective in inhibiting corrosion of steel. Conductive polymers have garnered much attention in recent years due to their environmentally benign nature and high effectiveness in protecting against corrosion. This book introduces the history of these conductive polymers. The applications of conducting polymers, polymer composites, and nanocomposites for corrosion protection of different industrial metal substrates are explored based on reported experimental data, and their mechanism of inhibition is explained. This book also includes overviews of some recent works on marine antifouling, marine heavy-protective coatings, and waterborne paints by conducting polymer and inorganic composites.

Conducting polymers and their family of stimuli-actuated polymers exhibit excellent application prospects in the fields of anticorrosion, antibacterial, anti-biofouling, and waterborne-based coatings. This book will be of great interest to students, scholars, and professionals alike in the field of corrosion engineering and material science.

**Dr. Yanhua Lei** received his doctoral degrees from the Department of Material Science and Engineering from Hokkaido University in 2014. He is now an associate professor at Shanghai Maritime University, China. His research interests include corrosion and protection, electrochemistry, and electro-catalysis, especially by using the conducting polymer and organic/inorganic composites.

# Corrosion and Protection of Marine Engineering Materials

## Applications of Conducting Polymers and Their Composites

Edited by
Yanhua Lei

CRC Press
Taylor & Francis Group
Boca Raton London New York

CRC Press is an imprint of the
Taylor & Francis Group, an **informa** business

Supported by Shanghai Collaborative Innovation Center of Heavy Icebreaker Key Steel, and Quality Evaluation and Foundation of Shanghai Engineering Technology Research Centre of Deep Offshore Material.

First edition published in English 2023
by CRC Press
6000 Broken Sound Parkway NW, Suite 300, Boca Raton, FL 33487-2742

and by CRC Press
4 Park Square, Milton Park, Abingdon, Oxon, OX14 4RN

CRC Press is an imprint of Taylor & Francis Group, LLC

English Version by permission of Shanghai Jiao Tong University Press.

Library of Congress Cataloging-in-Publication Data
Names: Lei, Yanhua, 1982- editor.
Title: Corrosion and protection of marine engineering materials : application of conducting polymers and their composites / edited by Yanhua Lei.
Description: First edition. | Boca Raton : CRC Press, 2023. | Includes bibliographical references. | Identifiers: LCCN 2022045805 (print) | LCCN 2022045806 (ebook) | ISBN 9781032452425 (hbk) | ISBN 9781032452722 (pbk) | ISBN 9781003376194 (ebk)
Subjects: LCSH: Ocean engineering--Materials--Protection. | Corrosion and anticorrosives.
Classification: LCC TC1650 .C67 2023  (print) | LCC TC1650  (ebook) | DDC 620/.4162--dc23/eng/20221128
LC record available at https://lccn.loc.gov/2022045805
LC ebook record available at https://lccn.loc.gov/2022045806

ISBN: 978-1-032-45242-5 (hbk)
ISBN: 978-1-032-45272-2 (pbk)
ISBN: 978-1-003-37619-4 (ebk)

DOI: 10.1201/9781003376194

Typeset in Minion
by MPS Limited, Dehradun

# Contents

# Preface

C orrosion is a process that takes place around us every day, and occurrences of corrosion are the result of a thermodynamically driven process of a metal converting to a more stable state. With the shortage of land resources, human beings have already turning to ocean; however, metals or alloys usually suffer from serious dissolution due to the complicated chemical-life environment of seawater. It has become a challenge to marine resources development, due to the serious corrosion and biofouling of marine engineering installation and transport equipment.

Coatings, as the most direct and effective method of anticorrosion, are widely used in various corrosion environments, especially in complicated marine conditions. However, with the requirements of environmental protection, there is thereby an urgent need, but it is still a significant challenge to develop new eco-friendly coatings with high performance.

The intrinsically conducting polymers have been a hot topic since their first appearance, attracting considerable attention from various areas, due to their easy synthesis, fewer toxicological problems, high electrical conductivity, good redox properties, as well as the processability of conventional polymers. The developments in nanomaterials and nanotechnologies largely expands the potential allocation of conducting polymers and their nanocomposites.

The conducting polymers, as a new corrosion-control methodology, have attracted more and more attention of researchers and related enterprises, owing to their excellent protective performance and environmental moderate features, especially under the requirement of sustainable development of the social environment.

In this book, conducting polymers are briefly introduced, including the history, synthesis, and conducting mechanism. Then, conducting polymers, especially based on conducting polymer coatings in corrosion protection, are reviewed, and several proposed protection mechanisms are introduced for understanding the protection of conducting polymers. Then, the works of the past ten years on conducting polymers of our group in corrosion protection, such as, steel, copper, zinc, and other different metal surfaces, are reviewed. Some recent works about the anticorrosion, antifouling, waterborne, and heavy-protective coatings by conducting polymers and conducting polymer/inorganic composites are also introduced in this book.

Due to my limited knowledge, there might be some mistakes and flaws in this book. Please don't hesitate to correct me.

**Yanhua Lei**
*College of Ocean Science and Engineering*
*Shanghai Maritime University*

# Acknowledgments

Throughout the writing of this book, I have received a great deal of support and assistance. I would particularly like to acknowledge my team members, Dr. Bochen Jiang, Mr. Da Huo, Mr. Mengchao Ding, and Dr. Zhangwei Guo for their wonderful collaboration and patient support.

I would also like to thank the support from Shanghai Collaborative Innovation Center of Heavy Icebreaker Key Steel, and Quality Evaluation and Foundation of Shanghai Engineering Technology Research Centre of Deep Offshore Material(19DZ2253100).

Acknowledgments for third-party material:

Table 1.1
Reprinted from Synthetic Metals, 273, Christopher Igwe Idumah, Novel trends in conductive polymeric nanocomposites, and bionanocomposites, Pages 116674, Mar 1, 2021, with permission from Elsevier.

Figure 1.7
Reprinted from Progress in Materials Science, 115, Xiao-Xiong Wang, Gui-Feng Yu, Jun Zhang, Miao Yu, Seeram Ramakrishna, Yun-Ze Long, Conductive polymer ultrafine fibers via electrospinning: Preparation, physical properties and applications, Pages 100704, Jan 1, 2021, with permission from Elsevier.

Figure 3.01
Reprinted from Arabian Journal of Chemistry, 13, A. Fateh, M. Aliofkhazraei, A.R. Rezvanian, Review of corrosive environments for copper and its corrosion inhibitors, Pages 481–544, Copyright (January 2020), with permission from Elsevier.

Figure 3.02
Reprinted from Elsevier Books, 12, P. Zhou, K. Ogle, Encyclopedia of Interfacial Chemistry, Pages 478–489, Copyright (Jan 1, 2018), with permission from Elsevier.

Figure 3.04, Figure 3.05
Reprinted from Corrosion Science, 46, G. Kear, B.D. Barker, F.C. Walsh, Electrochemical corrosion of unalloyed copper in chloride media--a critical review, Pages 109–135, Copyright (Jan 1, 2004), with permission from Elsevier.

Figure 3.10
Reprinted from Corrosion Science, 76, YanHua Lei, Nan Sheng, Atsushi Hyono, Mikito Ueda, Toshiaki Ohtsuka, Electrochemical synthesis of polypyrrole films on copper from phytic solution for corrosion protection, Pages 302–309, Copyright (Nov 1, 2013), with permission from Elsevier.

Figure 3.13
Reprinted from Progress in Organic Coatings, 77, YanHua Lei, Nan Sheng, Atsushi Hyono, Mikito Ueda, Toshiaki Ohtsuka, Influence of pH on the synthesis and properties of polypyrrole on copper from phytic acid solution for corrosion protection, Pages 774–784, Copyright (Apr 1, 2014), with permission from Elsevier.

Figure 3.18, 3.19, 3.20
Reprinted from Progress in Organic Coatings, 77, Y.H. Lei, N. Sheng, A. Hyono, M. Ueda, T. Ohtsuka, Effect of benzotriazole (BTA) addition on Polypyrrole film formation on copper and its corrosion protection, Pages 339–346, Copyright (Feb 1, 2014), with permission from Elsevier.

Table 4.01, 4.02
Reprinted from Progress in Organic Coatings, 76, Nan Sheng, Mikito Ueda, Toshiaki Ohtsuka, The formation of polypyrrole film on zinc-coated AZ91D alloy under constant current characterized by Raman spectroscopy, Pages 328–334, Copyright (February-March 2013), with permission from Elsevier.

Figure 4.1
Reprinted from Progress in Natural Science: Materials International, 26, Daokui Xu, En-hou Han, Yongbo Xu, Effect of long-period stacking ordered phase on microstructure, mechanical property and corrosion resistance of Mg alloys: A review, Pages 117–128, Copyright (April 2016), with permission from Elsevier.

Figure 4.2, 4.3, 4.10
Reprinted from Progress in Organic Coatings, 75, Nan Sheng, Toshiaki Ohtsuka, Preparation of conducting poly-pyrrole layer on zinc coated Mg alloy of AZ91D for corrosion protection, Pages 59–64, Copyright (September–October 2012), with permission from Elsevier.

Figure 4.11, 4.12, 4.13
Reprinted from Progress in Organic Coatings, 77, Nan Sheng, Yanhua Lei, Atsushi Hyonoo, Mikito Ueda, Toshiaki Ohtsuka, Improvement of polypyrrole films for corrosion protection of zinc-coated AZ91D alloy, Pages 1724–1734, Copyright (Nov 1, 2014), with permission from Elsevier.

Table 5.01
Reprinted from Progress in Materials Science, 104, Saviour A. Umoren, Moses M. Solomon, Protective polymeric films for industrial substrates: A critical review on past and recent applications with conducting polymers and polymer composites/nanocomposites, Pages 380–450, Copyright (July 1, 2019), with permission from Elsevier.

Figure 5.02
Reprinted from Progress in Organic Coatings, 114, G. Contri, G.M.O. Barra, S.D.A.S. Ramoa, C. Merlini, L.G. Ecco, F.S. Souza, A. Spinelli, Epoxy coating based on montmorillonite-polypyrrole: Electrical properties and prospective application on corrosion protection of steel, Pages 201–207, Copyright (Jan 1, 2018), with permission from Elsevier.

Figure 5.03, 5.04
Reprinted from Progress in Organic Coatings, 139, Yanhua Lei, Zhichao Qiu, Ning Tan, Hailiang Du, Dongdong Li, Jingrong Liu, Tao Liu, Weiguo Zhang, Xueting Chang, Polyaniline/CeO2 nanocomposites as corrosion inhibitors for improving the corrosive performance of epoxy

coating on carbon steel in 3.5% NaCl solution, Pages 105430, Copyright (Feb 1, 2020), with permission from Elsevier.

Figure 5.05
Reprinted from Materials Chemistry and Physics, 247, V.S. Sumi, S.R. Arunima, M.J. Deepa, M. Ameen Sha, A.H. Riyas, M.S. Meera, Viswanathan S.Saji, S.M.A. Shibli, PANI-Fe2O3 composite for enhancement of active life of alkyd resin coating for corrosion protection of steel, Pages 122881, Copyright (June 1, 2020), with permission from Elsevier.

Figure 5.06
Reprinted from Composites Part A: Applied Science and Manufacturing, 130, Qingsong Zhu, En Li, Xianhu Liu, Weiqiang Song, Mingyang Zhao, Lisen Zi, Xinchao Wang, Chuntai Liu, Synergistic effect of polypyrrole functionalized graphene oxide and zinc phosphate for enhanced anticorrosion performance of epoxy coatings, Pages 105752, Copyright (March 1, 2020), with permission from Elsevier.

Figure 5.09
Reprinted from Corrosion Science, 85, Sadegh Pour-Ali, Changiz Dehghanian, Ali Kosari, In situ synthesis of polyaniline–camphorsulfonate particles in an epoxy matrix for corrosion protection of mild steel in NaCl solution, Pages 204–214, Copyright (Aug 1, 2014), with permission from Elsevier.

Figure 5.11
Reprinted from Progress in Organic Coatings, 133, Xiangmiao Zhu, Zhongbin Ni, Liangliang Dong, Zhaokun Yang, Liming Cheng, Xiao Zhou, Yuxin Xing, Jie Wen, Mingqing Chen, In-situ modulation of interactions between polyaniline and graphene oxide films to develop waterborne epoxy anticorrosion coatings, Pages 106–116, Copyright (Aug 1, 2019), with permission from Elsevier.

Figure 5.12
Reprinted from Scientific Reports, 7, Haihua Wang, Huan Wen, Bin Hu, Guiqiang Fei, Yiding Shen, Liyu Sun, Dong Yang, Facile approach to fabricate waterborne polyaniline nanocomposites with environmental benignity and high physical properties, Pages 43694, Copyright (March 06, 2017), with permission from Springer Nature.

Figure 5.13
Reprinted from Applied Surface Science, 407, Shihui Qiu, Cheng Chen, Mingjun Cui, Wei Li, Haichao Zhao, Liping Wang, Corrosion protection performance of waterborne epoxy coatings containing self-doped polyaniline nanofiber, Pages 213–222, Copyright (June 15, 2017), with permission from Elsevier.

Figure 5.14
Reprinted from Acs Sustainable Chemistry & Engineering, 5, Kun Zhang and Surbhi Sharma, Site-Selective, Low-Loading, Au Nanoparticle–Polyaniline Hybrid Coatings with Enhanced Corrosion Resistance and Conductivity for Fuel Cells, Pages 277–286, Copyright (Jan 1, 2017), with permission from American Chemical Society.

Figure 6.02
Reprinted from Progress in Materials Science, 87, M.S. Selim, M.A. Shenashen, Sherif A. El-Safty, S.A. Higazy, M.M. Selim, H. Isago, A. Elmarakbi, Recent progress in marine foul-release polymeric nanocomposite coatings, Pages 1–32, Copyright (June 1, 2017), with permission from Elsevier.

Figure 6.04, 6.03A
Reprinted from Science of The Total Environment, 766, Liren Chen, Yanyi Duan, Mei Cui, Renliang Huang, Rongxin Su, Wei Qi, Zhimin He, Biomimetic surface coatings for marine antifouling: Natural antifoulants, synthetic polymers and surface microtopography, Pages 144469, Copyright (Apr 20, 2021), with permission from Elsevier.

Figure 6.05, 6.03B
Reprinted from Macromolecules, 42, Jenny A. Lichter, Krystyn J. Van Vliet, and Michael F. Rubner, Design of Antibacterial Surfaces and Interfaces: Polyelectrolyte Multilayers as a Multifunctional Platform, Pages 8573–8586, Copyright (October 12, 2009), with permission from American Chemical Society.

Figure 6.06
Reprinted from Colloids and Surfaces B: Biointerfaces, 150, L.A. Gallarato, L.E. Mulko, M.S. Dardanelli, C.A. Barbero, D.F. Acevedo, E.I. Yslas, Synergistic effect of polyaniline coverage and surface microstructure on

the inhibition of Pseudomonas aeruginosa biofilm formation, Pages 1–7, Copyright (Feb 1, 2017), with permission from Elsevier.

Figure 6.07
Reprinted from Biosensors and Bioelectronics, 135, Cai, Wei, Wang, Jixiao, Quan, Xiaodong, Wang, Zhi, Preparation of bromo-substituted polyaniline with excellent antibacterial activity, Pages 45657, Copyright (2017), with permission from John Wiley & Sons Books.

Figure 6.08
Reprinted from Biosensors and Bioelectronics, 92, Guixiang Wang, Rui Han, Xiaoli Su, Yinan Li, Guiyun Xu, Xiliang Luo, Zwitterionic peptide anchored to conducting polymer PEDOT for the development of antifouling and ultrasensitive electrochemical DNA sensor, Pages 396–401, Copyright (June 15, 2017), with permission from Elsevier.

Figure 6.09
Reprinted from Materials Science in Semiconductor Processing, 31, K. Pandiselvi, S. Thambidurai, Synthesis, characterization, and antimicrobial activity of chitosan-zinc oxide/polyaniline composites, Pages 573–581, Copyright (Mar 1, 2015), with permission from Elsevier.

Figure 6.10
Reprinted from Arabian Journal of Chemistry, 13, Ashraf P M, Sasikala K G, Thomas S N, Biofouling resistant polyethylene cage aquaculture nettings: A new approach using polyaniline and nano copper oxide, Pages 875–882, Copyright (January 2020), with permission from Elsevier.

Figure 6.12
Reprinted from Progress in Organic Coatings, 148, Changhao Wu, Jixiao Wang, Shuangshuang Song, Zhi Wang, Song Zhao, Antifouling and anticorrosion performance of the composite coating made of tetrabromobisphenol-A epoxy and polyaniline nanowires, Pages 105888, Copyright (Nov 1, 2020), with permission from Elsevier.

Figure 6.13
Reprinted from Applied Materials & Interfaces, 11, Jhih-Guang Wu, Jie-Hao Chen, Kuan-Ting Liu, and Shyh-Chyang Luo, Engineering Antifouling Conducting Polymers for Modern Biomedical Applications,

Pages 21294–21307, Copyright (May 23, 2019), with permission from American Chemical Society.

Figure 6.14
Reprinted from Langmuir, 35, Ahana Mohan and P. Muhamed Ashraf, Biofouling Control Using Nano Silicon Dioxide Reinforced Mixed-Charged Zwitterionic Hydrogel in Aquaculture Cage Nets, Pages 4328–4335, Copyright (February 28, 2019), with permission from American Chemical Society.

Figure 6.15
Reprinted from Materials Letters, 230, Meng-yang Jia, Jia-yi Zhang, Zhi-ming Zhang, Liang-min Yu, Jia Wang, The application of Ag@PPy composite coating in the cathodic polarization antifouling, Pages 283–288, Copyright (Nov 1, 2018), with permission from Elsevier.

Figure 6.16
Reprinted from Synthetic Metals, 257, Bo Wen, Geoffrey I.N. Waterhouse, Meng-Yang Jia, Xiao-hui Jiang, Zhi-Ming Zhang, Liang-min Yu, The feasibility of polyaniline-TiO2 coatings for photocathodic antifouling: antibacterial effect, Pages 116175, Copyright (November 2019), with permission from Elsevier.

# List of Abbreviations

**Chapter One: A General Introduction of Conducting Polymers in Corrosion Protection**

| | |
|---|---|
| **GDP** | Gross Domestic Product |
| **BTA** | Benzotriazole |
| **ICP** | Intrinsically Conducting Polymer |
| **PA** | Polyacetylene |
| **BLA** | Bond Length Alternation |
| **Eg** | Energy Gap |
| **CPs** | Conductive Polymers |
| **PPy** | Polypyrrole |
| **PTh** | Polythiophene |
| **PANI** | Polyaniline |
| **PEDTO** | Poly(3,4-ethylenedioxythiophene) |
| **PSS** | Polystyrene Sulfonate |
| **DoS** | Dodecyl Sulfate |
| **DBS** | Dodecyl Benzenesulfonate |
| **EQCM** | Electrochemical Quartz Crystal Microbalance |
| **pH** | Pouvoir Hydrogen |
| **DSM** | Ultra High Molecular Weight Polyethylene Fiber |
| **OTFTs** | Organic Thin-Film Transistors |
| **OLEDs** | Organic Light-Emitting Diodes |
| **OSCs** | Organic Solar Cells |

**Chapter Two: Corrosion Protection of Carbon Steel by Conducting Polypyrroles**

| | |
|---|---|
| **PMo** | $PMo_{12}O_{40}^{3-}$ |
| **PPy** | Polypyrrole |

**DoS**     Dodecyl Sulfate (In this chapter is $C_{12}H_{25}OSO_3^-$)
**CP**      Conducting Polymer
**PPy**     Polypyrrole
**Py**      Pyrrole
**PTHIO**   Polythiophene
**PANI**    Polyaniline
**CD**      Current Density
**SCC**     Stress Cracking Corrosion
**SEM**     Scanning Electron Microscopy
**NaDoS**   Sodium Dodecylsulfate

## Chapter Three: Inhibition of Copper Corrosion Using Conducting Polypyrrole Coatings

**Eqs**       Equations
**PPy**       Polypyrrole
**Py**        Pyrrole
**DBSA**      Dodecylbenzene Sulfonic Acid
**TOF-SIMS**  Time-of-Flight Secondary Ion Mass Spectrometry
**OXA**       Oxalic Acid
**EIS**       Electrochemical Impedance Spectroscopy
**OCP**       Open Circuit Potential
**PTh**       Polythiophene
**POT**       Poly(o-toluidine)
**POEA**      Poly (*o*-ethoxy aniline)
**PNEA**      Poly (*N*-ethyl aniline)
**EQCM**      Electrochemical Quartz Crystal Microbalance
**SEM**       Scanning electron Microscopy
**CD**        Current Density
**OCP**       Open Circuit Potential
**ICP-AES**   Inductively Coupled Plasma-Atomic Emission Spectrometry
**BTAH**      Benzotriazole

## Chapter Four: Corrosion Protection of Magnesium (Mg) Alloys by Conducting Polymers

**CPs**     Conductive Polymers
**PPy**     Polypyrrole
**PANI**    Polyaniline

PTh     Polythiophene
CD     Current Density
SEM     Scanning Electron Microscopy
OCP     Open Circuit Potential

## Chapter Five: Applications of Conducting Polymers for Anti-Corrosion in Marine Environments

| | |
|---|---|
| PANI | Polyaniline |
| PPy | Polypyrrole |
| PTh | Polythiophene |
| CPs | Conductive Polymers |
| WEP | Waterborne Epoxy Coatings |
| WBE | Wire Beam Electrode |
| PEDOT | Poly(3,4-ethylenedixythiophene) |
| PPD | Poly(phenylenediamine) |
| Mt | Montmorillonite |
| Cdl | Double Layer Capacitance |
| Rp | Polarization Resistance |
| EIS | Electrochemical Impedance Spectroscopy |
| IPN | Interpenetrating Polymer Network |
| NPs | Nanocomposite |
| TEM | Transmission Electron Microscope |
| HRTEM | High Resolution Transmission Electron Microscope |
| CSA | Camphor Sulfonic Acid |
| CNTs | Carbon Nanotubes |
| ZRP | Zinc-Rich Epoxy Primer |
| CSA | Camphor Sulfonic Acid |
| Gr | Graphene |
| OCP | Open Circuit Potential |
| ZRC | Zinc-Rich Epoxy Resin |
| PEO | Polyethylene Oxide |
| AP-g-PEO | Polyethylene Oxide-grafted 3-aminophenol |
| PVA | Polyvinyl Alcohol |
| PAA | Polyacrylic Acid |
| COWPU | Castor Oil-based Waterborne Polyurethane |
| SPANI | Sulfonated Polyaniline |
| CPANI | Carboxylate Polyaniline |
| PEMFC | Proton Exchange Membrane Fuel Cell |

| | |
|---|---|
| **EP** | Epoxy Coatings |
| **WEP** | Waterborne Epoxy Coatings |
| **VOCs** | Volatile Organic Compounds |
| **PEMFC** | Proton Exchange Membrane Fuel Cell |

## Chapter Six: An Introduction of Conducting Polymers in Anti-Bacterial and Anti-Biofouling Applications

| | |
|---|---|
| **MIC** | Microbiologically Influenced Corrosion |
| **PANI** | Polyaniline |
| **SRB** | Sulfate-Reducing Bacteria |
| **EPS** | Extracellular Polymeric Substances |
| **IOB** | Aerobic Iron-Oxidizing Bacteria |
| **EPS** | Extracellular Polymeric Substances |
| **ZnPT** | Zinc Pyrithione |
| **PDMS** | Polydimethylsiloxane |
| **PTFE** | Poly Tetra Fluoroethylene |
| **TBT** | Tributyltin moiety |
| **CPs** | Conducting Polymers |
| **PANI** | Polyaniline |
| **PPy** | Polypyrrole |
| **SPAN** | Sulphonated Derivative |
| **PANIEB** | Polyaniline Emeraldine Base |
| **PANI-Cl** | Polyaniline Chloride |
| **Gr** | Graphene |
| **PA** | Phytic Acid |
| **DLIP** | Direct Laser Interference Pattern |
| **Br-PANI** | Brominated Polyaniline |
| **PET** | Polyethylene Terephthalate |
| **PES** | Polyethersulfone |
| **UF** | Ultrafiltration |
| **CLSM** | Confocal Laser Sanning Mcroscopy |
| **DCE** | 1.2-dichloroethane |
| **EB** | Emeraldine Base |
| **DBSA** | Dodecyl Benzenesulfonic Acid |
| **ROS** | Reactive Oxygen Species |

# Contributors

**Qing Chen**
School of Material Science and
　Engineering
Tongji University
Shanghai, China

**Bochen Jiang**
College of Ocean Science
　and Engineering
Shanghai Maritime University
Shanghai, China

**Yanhua Lei**
College of Ocean Science and
　Engineering
Shanghai Maritime University
Shanghai, China

**Tao Liu**
College of Ocean Science
　and Engineering
Shanghai Maritime University
Shanghai, China

**Toshiaki Ohtsuka**
Faculty of engineering
Hokkaido University
Sapporo, Hokkaido, Japan

**Nan Sheng**
China University of Mining
　and Technology
Xuzhou, China

**Jingxiang Xu**
College of Engineering Science
　and Technology
Shanghai Ocean University
Shanghai, China

# CHAPTER 1

# A General Introduction of Conducting Polymers in Corrosion Protection

Yanhua Lei and Jingxiang Xu

## CONTENTS

DOI: 10.1201/9781003376194-1

## 1.1  BACKGROUND OF CORROSION

Corrosion is a process that takes place around us every day. Its aspects are commonly recognized in everyday life and range from squeaking hinges, nuts and bolts that cannot be loosened, to the brittle red/brown material that's eating away at our almost "brand new" car. The occurrences of corrosion are the result of a thermodynamically driven process of a metal converting to a more stable state. Usually, this state is a metal oxide, a similar state to the material commonly found in nature from which the metal was refined from in the first place. For this reason, the processes of corrosion has been described as metallurgy in reverse.

The economy of corrosion is the basis of significant cost to the nations. Corrosion concerns almost every industry and sector in the country. Corrosion will cause structural damage to facilities and equipment, shorten service life, and more seriously, sometimes serious corrosion may cause sudden disaster accidents. The economy of corrosion is the basis of significant cost to nations. For example, it was reported that the total cost of corrosion in the United States for the years 1999–2001 was a staggering total of $276 billion. This is approximately 3% of the Gross National Product of the period of study.[1] A report released by NACE International in 2016 estimated that the global cost of corrosion was approximately USD$2.5 trillion, excluding the cost of corrosion failure consequences on the safety and the environment. However, the report also stated that these damage costs could be reduced to 15% and 35% by implementing excellent control and management practice, respectively. In 2015, the Chinese Academy of Engineering set up a major consulting project on "Corrosion Status and Control Strategy Research of China", which was under the charge of Professor Baorong Hou. The latest corrosion investigation results show that the losses and anti-corrosion investments caused by corrosion in China accounted for about 3.34% of the GDP of 2014, with the total amount exceeding 2 trillion yuan.

What is worse is that corrosion of metal structures often lead to human injuries and death.[2] For example, due to the combination of stress and corrosion, the collapses of the "Silver Bridge" over the Ohio River linking

Point Pleasant carried 46 people to their deaths on December 15, 1976.[1] On August 2013 in Japan, 300 metric tons of contaminated water leaked from a storage tank, a disaster that was described by Japan's nuclear regulator as the worst accident at Fukushima since the earthquake and tsunami of 2011 caused reactors to melt.[3] The cause of the failure was traced to corrosion around faulty seals. In the same year, in Huangdao, Shandong Province, due to serious corrosion of an oil pipeline, the serious leakage oils resulted in a devastating explosion, which caused 62 deaths, 136 injuries, and a direct economic loss of 750 million RMB. Investigations revealed that the direct cause of the explosion was the ignition of vapors that originated from oil leaked from a corroded underground pipeline when workers used a non-explosion-proof hydraulic hammer.[4]

Corrosion is a safety problem, an economic problem, and an ecological civilization problem. Corrosion protection is an important part of the development of the "One Belt And One Road" strategy, and the prevention and control of corrosion is a reflection of national civilization and prosperity. Compared with developed countries such as Europe and the United States, China's corrosion investigation started late, and its coverage field is preferred. Therefore, it is urgent to carry out corrosion cost investigations and anti-corrosion strategy research at the national level, so as to strengthen people's understanding of corrosion hazards and improve people's anti-corrosion awareness.

Since nothing can be done to alter the thermodynamics of the corrosion process, corrosion control strategies focus on controlling the dynamics (slowing the kinetics and/or altering the mechanism) of the process.[5] It is believed that implementing the best metals' corrosion prevention measures could result in global savings of 15–35% of the cost spent on corrosion annually.[3]

Various strategies have been used to combat corrosion. These can be grouped into five types: (i) design, (ii) materials selection, (iii) electrochemical (anodic and cathodic protection), (iv) coatings, and (v) modification of environment (the use of corrosion inhibitors). The design strategy involves designing a material or system in a way to avoid crevices and excessive velocities or localized turbulence.[1]

The pursuit of means to delay the process of corrosion has led to today's worldwide use of corrosion inhibition, such as chromates and benzotriazole (BTA). However, due to the toxicity and environmental effects, the use of chromates has been restrained legally.[6,7] Further, coatings, as the most direct and effective method of anti-corrosion, are widely used in metal corrosion protection. However, with the requirements of environmental protection, the need of new eco-friendly materials with high performance in corrosion protection have been the subject of many studies. The intrinsically conducting polymer (ICP), as a new corrosion-control methodology, have attracted more and more attention from researchers and related enterprises.

## 1.2 INTRODUCTION TO INTRINSICALLY CONDUCTING POLYMERS

### 1.2.1 Discovery of Intrinsically Conducting Polymers

Most of the commercially produced organic polymers are electric insulators. The ground-breaking accomplishments of Shirakawa, MacDiarmid, and Heeger in 1977 have fundamentally changed our view of organic polymers, from insulating 'plastics' to electrically (semi)conducting functional materials.[8] Starting from the serendipitous synthesis by Shirakawa's group of a bright and silvery polymer, trans-polyacetylene, collaborative efforts with MacDiarmid's and Heeger's teams realized the great potential of this material. The discovered polyacetylene (PA) could be doped to the metallic state, and in this stage it could exhibit electrical conductivity comparable to cupper, iron, and silver, and initiated investigations of a new class of polymeric materials that are called "synthetic metals" or "intrinsically conductive polymers" (ICPs). The conducting of PA was observed to increase from a semi-conductor to a metallic range with a conductivity of up to $10^5$ S/cm after partial oxidation of the polymer by exposure to halogen vapours (Figure.1.1a).

Before the discovery of PA, the CP prototype was the inorganic material poly(sulfur nitride) $(SN)_x$, a material exhibiting intrinsic (thus without doping) metallic conductivity where electrons in a partially filled conduction band move freely at long (ballistic) distances. In $(SN)_x$, all bonds have equal length, meaning that the bond length alternation (BLA) along the chain is zero, a key feature to achieve high conductivity.[8] However, it has been suggested that the structure of PA is more energetically stable when the chain has a BLA $\neq 0$, which opens an energy gap (Eg) between the top of the valence band and bottom of the conduction band, resulting in insulating behavior. The effect of doping a polymer—for instance, by exposing it to halogen vapors, as done by the three Nobel Laureates, is to reduce the energy gap and enhance the

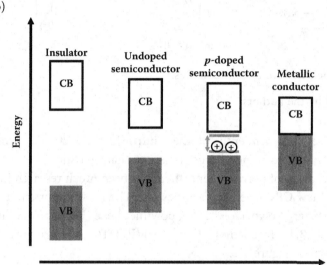

FIGURE 1.1    Structures, synthesis, and properties of CPs. (a) Chemical structure and trans-polyacetylene as well as doping of the latter polymer with an elemental halogen (for example, $Br_2$); (b) Schematic evolution of the band structure going from an insulator to a semiconductor to a conductor. Note, the effect of doping is to decrease the bandgap (Eg), increase charge carrier density, and enhance electrical conductivity.

**Source: created by the authors.**

carrier density to achieve a different degree of semiconducting (or even metallic) behavior[8] (Figure 1.1b).

This discovery laid the foundation for a whole range of novel materials possessing electrical conductivity; at the same time, the discovery brought them the Nobel Prize in the year 2000.[9] Since these pioneering studies, the field of CPs took off, with joint efforts from chemists, materials scientists, and physicists at universities and in industries. Note, with CPs we will here refer to any π-conjugated polymer — polymer having a backbone with alternating single and double (or triple) covalent bonds — that can

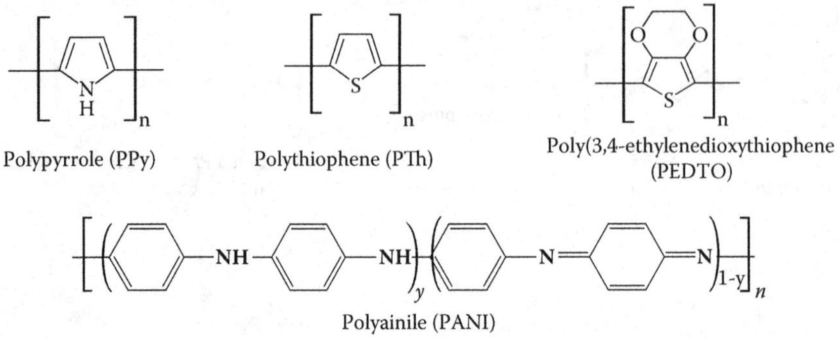

Polypyrrole (PPy)          Polythiophene (PTh)          Poly(3,4-ethylenedioxythiophene (PEDTO)

Polyainile (PANI)

FIGURE 1.2   The structure of common CPs.

**Source: created by the authors.**

transport charges, independently of their intrinsic conductivities (conductor or semiconductor) and charge transport characteristics.

Since the discovery of polyacetylene, there has been much research into CPs and many new CPs have been synthesized. The most important, and common, of these are polypyrrole (PPy), polythiophene (PTh), polyaniline (PANI), and poly(3,4-ethylenedioxythiophene) (PEDTO), the structures of which are shown in Figure 1.2.

## 1.3 INTRODUCTION TO SEVERAL CPS

### 1.3.1 Polyaniline (PANI)

PANI is an organic conducting polymeric material that captured research interest after the discovery of conjugated polyacetylene, due to the advantages such as ease of synthesis process, low-price, excellent polymerization yield, as well as high electrical conductivity and redox activity.[10,11]

PANI exists in a variety of forms that differ in chemical and physical properties. It was found that polyaniline consists of alternating reduced and oxidized repeating units (Figure 1.3). The fully oxidized form, the fully reduced form, and the half-oxidized forms are known as the leucoemaraldine base form (LEB), pernigraniline form (PNB), and emaraldine base form (EB), respectively.

The most common form, green protonated emeraldine (Figure. 1.3c), has a conductivity on a semiconductor level in the order of 100 S cm$^{-1}$, many orders of magnitude higher than that of common polymers ($<10^{-9}$ S cm–1) but lower than that of typical metals ($>10^4$ S cm$^{-1}$). Protonated PANI, is also named PANI emeraldine salt (PANI-ES), e.g., PANI hydrochloride,

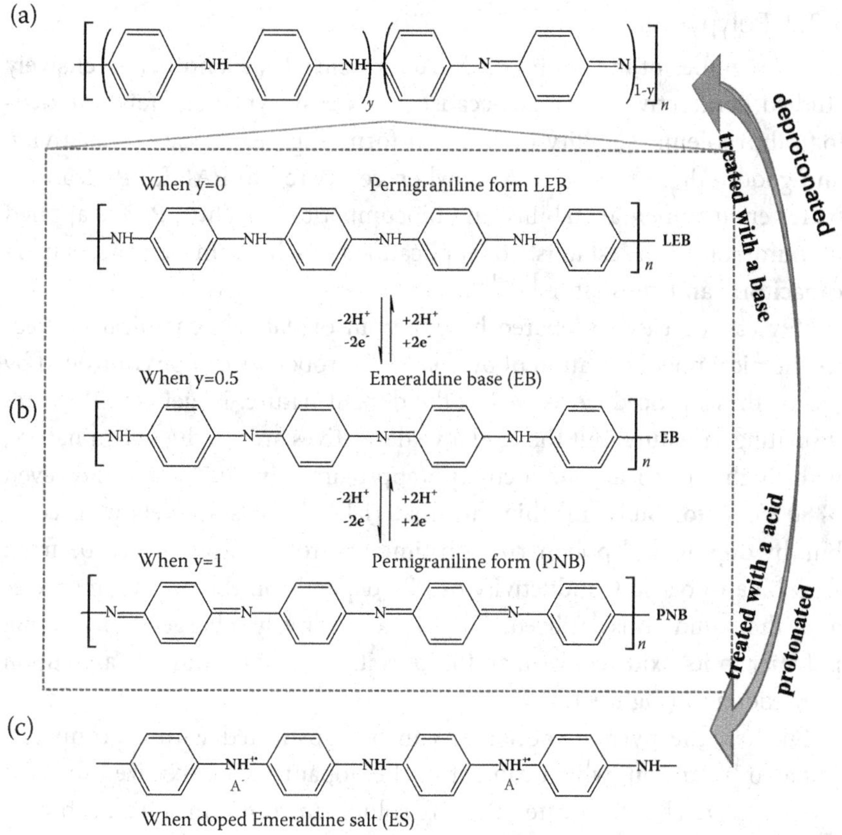

(a)

(b)

When y=0    Pernigraniline form LEB

LEB

$-2H^+$ | $+2H^+$
$-2e^-$ | $+2e^-$

When y=0.5    Emeraldine base (EB)

EB

$-2H^+$ | $+2H^+$
$-2e^-$ | $+2e^-$

When y=1    Pernigraniline form (PNB)

PNB

deprotonated
treated with a base
treated with a acid
protonated

(c)

When doped Emeraldine salt (ES)

FIGURE 1.3 Molecular structure of (a) polyaniline and (b) three oxide state of polyaniline in full oxidized form (PNB), the fully reduced form (LEB), and the half-oxidized form (EB); (c) protonated (doped) form of polyaniline (emaraldine). Polyaniline (emeraldine) salt (ES) is deprotonated in alkaline medium to polyaniline (emeraldine) base. $A^-$ is an anion.

Source: created by the authors.

converts to the non-conducting blue emeraldine base when treated with a base (Figure 1.3, such as ammonium hydroxide).

PANI is considered one of the CPs most used as an anti-corrosion coating. It has the following advantages over other conducting polymers: (1) it can be synthesized easily via chemical or electrochemical means, (2) it can be doped and dedoped easily by treatment with aqueous acid and base, (3) it cannot be degraded easily, and (4) the aniline monomer is relatively cheap.

## 1.3.2 Polypyrrole

PPy, a member of the conjugated heterocyclic CPs, is another extensively studied conductive polymer because of its easy synthesis, fewer toxicological problems, stability in oxidized form, high electrical conductivity, and good redox properties. And, when compared to PANI, PPy features better environmental stability and biocompatibility. Thus, PPy is applied in numerous well-established applications, such as in sensors, supercapacitors, and resonators[12–14].

PPy can be easily prepared by either an oxidatively chemical or electrochemical polymerization of pyrrole. The properties can be controlled by the synthesis procedure as well as the dopant nature. In general, PPys are insulating in nature, but the oxidative derivatives are conductive in nature, with high electrical conductivity (approximately 105 S/cm and even >380 S/cm) for bulk and thin-film materials[15]. PPys are yellow in color but, by doping and polymerization time control, their color can be from dark blue to black. Conductivity of PPy depends on the oxidation species and the conditions utilized. PPy is a positively charged conducting polymer in its oxidized form and it loses its conductivity and charge upon overoxidation (Figure 1.4).

Further, the pyrrole monomer can be substituted with a group terminated by an ion, which can act as the dopant in the oxidized form of the polymer. The substituted PPy are soluble in common solvents, but an increase in solvent processability is always accompanied by a loss of conductivity.

Due to the advantages described as well as PANI, PPy is another most promising materials used in corrosion protection. Various protection mechanisms have been proposed for the PPy-coated metals and alloys, including ennobling or anodic passivation, cathodic protection, mediation of oxygen reduction, barrier protection, and controlled inhibitor release mechanism.

## 1.3.3 Polythiophene

Polythiophene (PTh) is similar to PPy in structure; the only difference is the presence of the heteroatom "S" in the place of "N" atom in the aromatic ring. The PTh polymer gained significant attention in research and industrial areas because it possesses high environmental stability, better thermal stability, and less bandgap energy. Moreover, their interesting properties like semiconducting, electronic and optical activities,

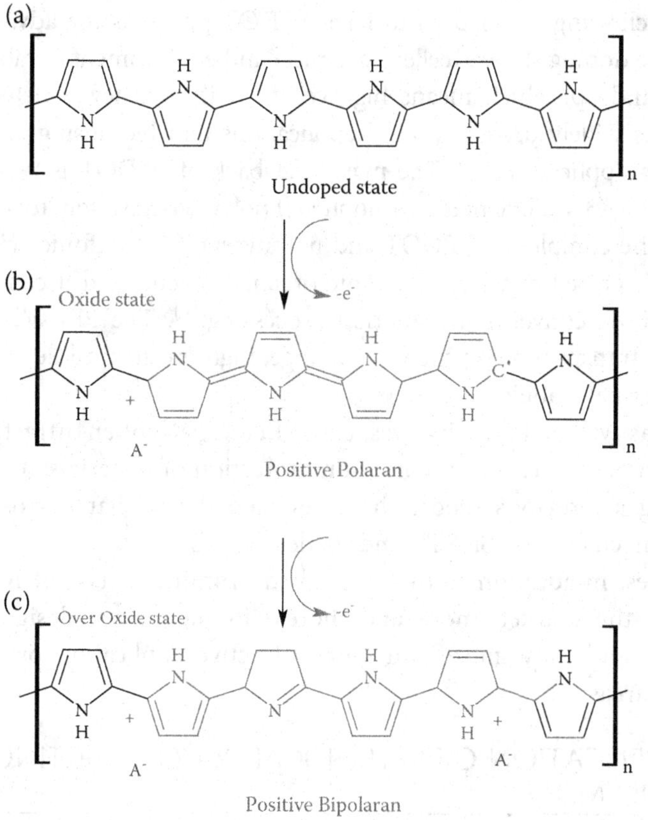

FIGURE 1.4 Molecular structure of (a) olypyrrole undoped state, (b) olypyrrole in oxide state, (c) polypyrrole in over oxide state.

Source: created by the authors.

along with their better mechanical characteristics and ease of process-ability brought significant attention to the PTh composites[16]. Similarly, in the case of other PPy, the PTh matrix can also be changed to a more conducting one by creating polarons and bipolarons in the backbone through either oxidation or reduction.

Unfortunately, similar to other reported CPs, the main drawback of PTh remains the lack of solubility due to its strong interchain interactions, resulting in limited processability. Water-soluble, stable, and highly con-ducting PTh derivatives were prepared by self-acid-doping with sulfonic acid. Due to poor solubility and processability of PTh, a great part of studies have focused on substituted PTh derivatives such as PEDOT, which has dioxyalkylene bridging group across the 3- and 4-positions of its heterocyclic ring. Comapred to PTh, PEDOT proceeds the advantages of reversible doping state, excellent chemical and environmental stability, and a low bandgap, which means high conductivity and fewer defects[17–19]; thus, it is widely used in optical applications and biosensing and bioengineering applications[20]. The main drawback of PEDOT is its poor solubility. Poly(3,4-ethylenedioxythiophene) poly(styrenesulfonate) (PEDOT-PSS) is the complex of PEDOT and polystyrene (PS) sulfonic acid, which can be dispersed in water and some organic solvents and it can be easily prepared via conventional solution processing[16]. The PEDOT-PSS film has high transparency in the visible range, high mechanical flexibility, and good thermal stability.

PTh as well as its derivatives, e.g., PEDOT, are often attracted much attention of researcher for corrosion protection of materials, due to their advantages. Previous reports have revealed the similar corrosion protective mechanisms for PTh and its derivatives[21–29].

Besides, in addition to the mentioned prominent class of ICPs presently in the market, more and more CPs have been designed, constructed, and show more and more attractive application prospects in various areas.

## 1.4 APPLICATION OF INTRINSICALLY CONDUCTING POLYMER

CPs have economic importance because of their functional elements, good environmental stability, flexibility, and electrical conductivity as well because of their useful mechanical, optical, and electronic properties[30]. CP applications depend on their processing characteristics, the doping (charge density) level, redox properties, and whether the charge transport is purely electronic or mixed ionic/electronic. The first CP use was in areas requiring heavily doped/highly conducting materials. Importantly, it was observed that objects placed in the reaction mixture for the electrochemical synthesis of PANI or PPy became preferentially coated with a highly CP film; this opened the way to explore using CPs as electrode materials in batteries and in antistatic/anticorrosion coating layers. However, commercial success in these areas has varied greatly: while their use as antistatic/electromagnetic shielding has found extensive commercialization, with the CP market reaching ~USD$900 million in 2018, commercial batteries based on CPs are unlikely to materialize[31]. Although recent studies suggested CP-based lithium ion batteries with performance and stability comparable/surpassing those of inorganics, CP cost is a roadblock, considering the large quantity of electro-active CP required for their fabrication (>100–1,000 times larger than the need for other opto-electronic devices)[8]. More recent applications are in organic solar cells (OSCs) for photovoltaic modules, organic thin-film transistors (OTFTs) for displays and circuits, and organic light-emitting diodes (OLEDs) for display and lighting. For OSC modules, which are expected to be a ~USD$150 million market in 2022 (of the ~USD$200 billion photovoltaic market), CPs can potentially be used in several of the device components where either a heavily doped or undoped CP film is needed. For instance, PEDOT:PSS (Figure 1.2), formulated in different ways to tune work-function, conductivity, and rheological properties, has been widely investigated as an interfacial hole transport layer, as well as, with less success, to replace the conducting transparent electrode[8].

In CP nanocomposites, polymers function as matrices and different nanofillers embedded in the matrix result in the formation of new material that can be used in various applications as per requirements. During the past decades, there have been many reports concerning investigations leading to practical applications of CPs in various fields, such as charge storage devices and batteries,[12,32–34] matrices for catalysis, [35–40] supercapacitors, [41–45] sensors, [46–50] and corrosion inhibitors.[51–56] An overview of CPs and CP-based nanocomposites and applications is illustrated in Figure 1.5.

## 1.5 CONDUCTIVITY MECHANISM OF ICPS

Conventional polymers such as epoxy resin can be modified by blending polymers with pigments to improve electrical, mechanical, or anti-corrosion properties. For example, electric conductivity of conventional polymers may be improved by introducing carbon or metallic particles as pigments, leading to application to anti-static coatings. In this case, electrically conductive pigments play a role of electric lead and are allowed to reduce or prohibit a static-electrical charge. In this case, the conductivity is not ascribed to the structure of polymers as distinguished from CPs. When Shirakawa discovered the method to prepare a free-standing film of PA,[4] many theoretical models were proposed to predict the conductivity of PAs. The undoped PA of cis-PA was an insulator at $\sigma = 1.7 \times 10^{-9}$ S/cm and trans-PA was a low conductivity at $\sigma = 4.4 \times 10^{-5}$ S/cm. The PA with a high conductivity was realized by MacDiarmid and Heeger, who discovered the method of how to chemically dope fibrillar crystalline PA.[9]

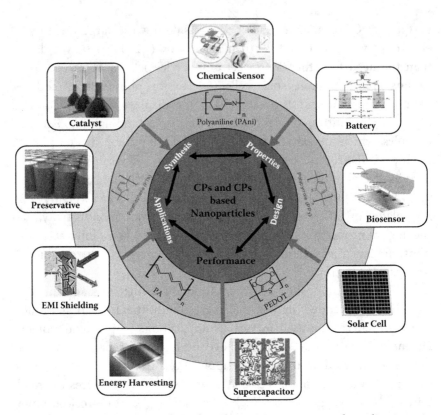

FIGURE 1.5  An overview of CP-based nanocomposites and applications in various areas.

**Source: created by the authors.**

The conductivity increase with the doping of CPs expanded the application field.

As a class of materials, CPs share several characteristics, including macromolecular character and electrical transport properties. The common structure in all ICPs is an altering single and double bond along the polymer backbone. As in the case with PAs, the simplest CP structure, the electrical transport characteristics are obtained by placing the alternating carbon-carbon double bond structure directly on the backbone for the polymer. The subsequent electrical conductivity if often mistakenly visualized from resonance structures and the mobility of π electrons as depicted for PAs in Figure 1.6a.

Compared to the structure in (a), the doped structure successfully transports charge across the polymer chain. Conjugation alone, however,

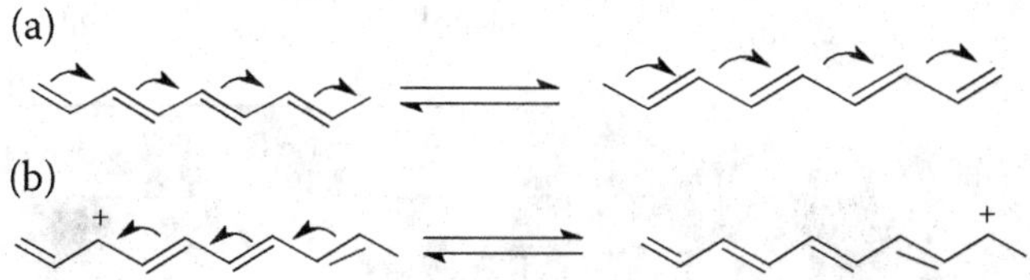

(a)

(b)

FIGURE 1.6 (a) Polyacetylene structure and resonance of electronic states mistakenly attributed to conductivity; (b) polyacetylene showing the effect of dopant in conductivity.

**Source: created by the authors.**

is not sufficient for conductivity. The charge carriers (electrons and holes) must be provided extrinsically by a doping process. A dopant is required to alter the band structure of the semiconductive polymer backbone. The defects caused by the dopant allow electrical transport either by hole or electron mobility, described in Figure 1.6. To make polymers more conductive, the charge carriers or "holes" have to be introduced into a polymer backbone. The p-type doping process to conjugated polymers, from which electrons are removed, and the n-type doping process to the polymers, from which electrons are added, was made by a reaction with oxidants and reductants, respectively. The doping of CPs involves the addition of a very high concentration of doping agent, in contrast to substitution with doping impurity at a low concentration in semiconductors, such as silicon and germanium. In a conducting polymeric system, the polymer must possess both overlapped molecular orbital which allows charge carries (Figure 1.7).

Since the discovery of conducting PAs, a variety of CPs have been developed and widely used due to relatively high conductivity, as listed in Table 1.1.

## 1.6 SYNTHESIS OF INTRINSICALLY CONDUCTING POLYMERS

CPs are usually synthesized via oxidative coupling of monomers. For polymerization, the first step is the oxidation of the monomer, which results in the formation of a radical cation, which then reacts with another monomer or radical cation, forming a dimer. Hence, an obvious classification is the initiation process of polymerization. The three general initiation routes are chemical, electrochemical, and photo-induced polymerization, each with its own advantages and disadvantages.

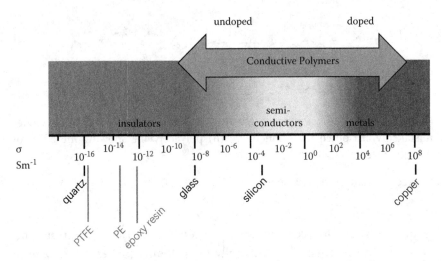

FIGURE 1.7 The electric conductivity of CPs and conventional polymers.

**Source: Progress in Materials Science (Elsevier).**

TABLE 1.1 Widely Used CPs with High Conductivity

| Conducting Polymers | Conductivity (S cm$^{-1}$) | Type of doping |
|---|---|---|
| Polyaniline (PANI) | 30–200 | n-type/p-type |
| Polypyrrole (PPy) | 10–7,500 | p-type |
| Poly(3,4-ethylenediox-ythiophene) (PEDOT) | 0.4–400 | n-type/p-type |
| Polythiophene (PT) | 10–1,000 | p-type |
| Polyparaphenylene (PPP) | 500 | n-type/p-type |
| Polyparaphenelene sulfide (PPS) | 3–300 | p-type |
| Polyisothionaphthene (PITN) | 1–50 | p-type |

*Source:* Science China Materials (Elsevier).

In chemical polymerization, chemical oxidants (such as ferric chloride or ammonium persulfate) are applied to oxidize the monomer. The morphology of the polymer synthesized by a chemical route can be controlled by varying the parameters of the process, such as monomer/oxidant concentration, temperature, pH, and reaction time. In chemical polymerization, a chemical oxidant is used to oxidize the monomer as well as provide a counter ion to function as a dopant, shown for the polymerization of pyrrole in Figure 1.8.

This approach to producing CPs is extensively used in industries (e.g., by DSM, Mitsubishi Rayon, Ormecon Chemie), but is limited to the small number of oxidants that can both oxidize the monomer and provide a suitable dopant. Chemical oxidation most often results in the

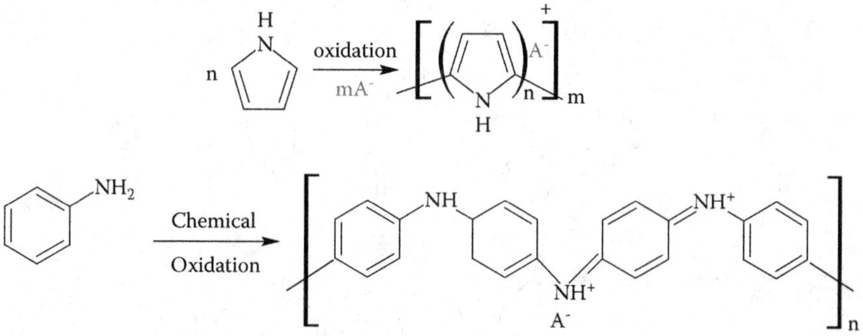

FIGURE 1.8  Schematic representation of polymerization of pyrrole and aniline, respectively to give PPy and PANI with the incorporation of a dopant anions A-.

**Source: created by the authors.**

formation of a CP powder, which generally displays lower conductivity than electrochemically prepared CPs.

Electro-polymerization for synthesis of CPs is preferred because it effectively controls the polymerization process and deposits the polymer on the surface of a metal substrate. It easily controls coating parameters. Most CPs can be electrochemically produced by anodic oxidation. The polymerization potential of CPs is essential from the viewpoint of electrochemistry. A large difference between oxidation potential from the monomer to polymerization and low corrosion potential of less noble metals may introduce a problem. The potential difference inhibits polymerization-deposition of CPs on less noble metals. PPy possesses relatively low oxidative-polymerization potential and, therefore, is widely investigated for the corrosion protection of the less noble metals. In comparison to PANI, PPy is further physiochemically stable in a wide pH range of aqueous solutions.

There have been many reports concerning the electro-polymerization process of CPs on metals like iron, carbon steel, or stainless steel.[15,30,55,57–60] Herein, an example of PPy was selected to illustrate the process of electro-polymerization as follows.

During the oxidizing electro-polymerization of Py, two reactions take place on the electrode: i) oxidation of Py monomer, and ii) oxidation of polymer backbone, producing positive charges. The positive charges are compensated by the anions doped from electrolytes. This reaction is identified as a doping process. Figure 1.9 shows the electro-polymerization reaction of pyrrole (1) in which a polymer film is formed and deposited on

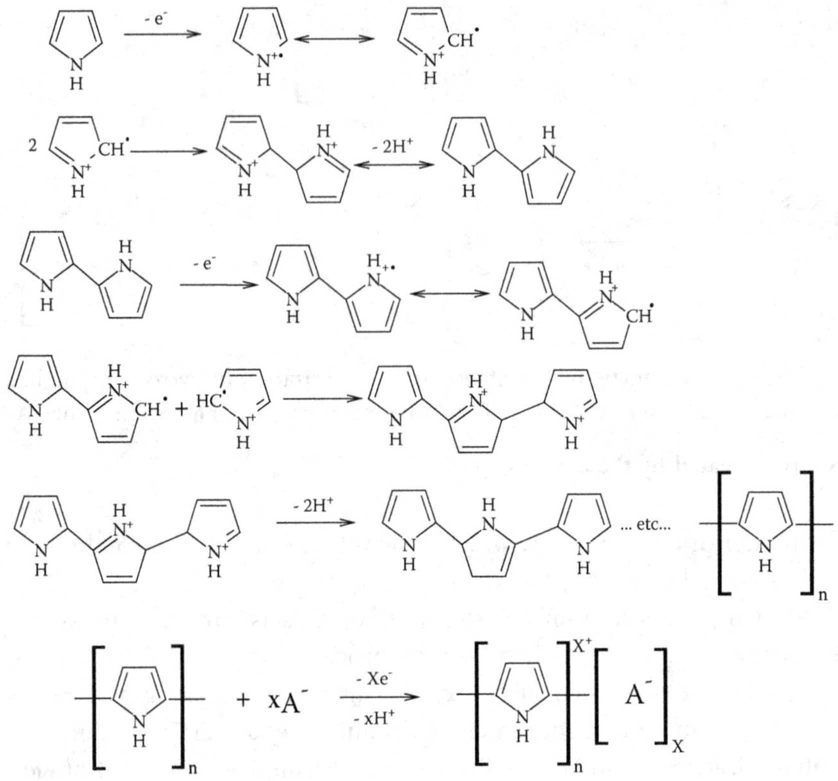

FIGURE 1.9   The mechanism of the polypyrrole polymerization.

**Source: created by the authors.**

an electrode surface. Reaction (2) represents the simultaneous removal of the electron and doping of counter-anions.

PPy can be obtained on a substrate surface by potentiodynamic, potentiostatic, and galvanostatic polymerization, or modifications of those methods. Excluding the passive film formation, the potentiostatic and galvanostatic methods have been widely applied. In the potentiostatic method, optimum voltage is applied during synthesis. In the galvanostatic method, a fixed oxidation current is supplied without control of the resulting potential; however, it allows easy control of the thickness of the deposited CP film by the charge passed.

## 1.7 APPLICATION OF ICP IN CORROSION PROTECTION

Conductive PANI was first reported to provide passivation to stainless steel under highly acidic conditions in 1985 by Deberry,[61] who found that the stainless steel covered by PANI was kept in the passive state for relatively long period in a sulfuric acid solution. This finding opened a new door to protect metal against corrosion. This new corrosion-control methodology has a potential for replacing the traditional hexavalent chromium technology. Since then, the field of corrosion protection using ICPs has generated significant interest in the last 30 years.

Coatings on the surface of metals by polymeric materials have been widely used in industries for the protection of these materials against corrosion.[62–65] Some specific ICPs like PANI, PPy, and their derivatives, have been found to display interesting corrosion protection properties. The effective use of CPs for corrosion protection of metals can be carried out by the following methods: (1) being used as pigments in various paints; (2) being directly electrodeposited on metal surface as primer; (3) being as corrosion inhibitors a in the corrosive solution.

## 1.7.1 CP-Blended Paint/Resin

Alternatives to casting-based methods are to disperse the conducting polymer into a paint or into epoxy or acrylic resins. With this approach, finding a suitable solvent can be bypassed since the conducting polymer can be dispersed in its solid form. In addition, it is possible with this approach to achieve the synergistic effect of the mechanical properties of the paints or resins and the electrical properties of the conducting polymers. This is also the method of choice in expanding the application of conducting polymers to the industrial level, especially in coating large structures.[66]

## 1.7.2 CP Electropolymerized/Electrodeposited Coating

As previous described in section 1.6, CPs can easily be synthesized on the conducting substrates from monomer-dispersed electrolyte solution with electrochemical methods. Electrochemical polymerization does not require the use of oxidant. It can be synthesized either in two or three electrode cells.

Compared to the chemical oxidation methods to synthesis CPs, an electrochemistry route is a straightforward means of obtaining CP films with a certain thickness by controlling the number of cycles or the current that is applied to the electrode. The CP coatings can be selectively and precisely deposited, which is a patterned film preparation technology. Polymeric films can be selectively and precisely deposited, which is a patterned film preparation technology. In addition, the doped anions can be carefully selected from the electrolyte during the electrodepositing process. This is usually the method of choice in coating relatively small areas. Care must be taken in choosing the electropolymerization condition, especially the applied potential and current. The applied potential should be high enough to oxidize the monomer and polymerize it, but low enough not to dissolve the metal or induce corrosion.

### 1.7.3 CP-Based Inhibitors

The use of corrosion inhibitors is the one the most economical and practical means of controlling metallic corrosion in different corrosive media. A metal corrosion inhibitor is a substance with a capacity to appreciably suppress corrosion when added in a minute amount to a corrosive medium.[3] The inorganic metals corrosion inhibitors function primarily by oxidizing a metal surface to form an impervious oxide layer, while the organic counterparts are inhibited by an adsorption mechanism through the use of their heteroatoms and/or unsaturated bonds as the adsorption center.[3,67]

Solubility is an essential requirement for corrosion inhibition, especially for the polymer. In spite of the fact, many CP molecules satisfy the conditions demanded as corrosion inhibitors, such as possessing functional groups containing hetero-atoms (N, O, S, P, etc.) that can donate their lone pairs of electrons, and always owning the ability to form a passive/protective layer on metal surface.[3,68] However, a search of the corrosion literatures revealed that research work on CPs as metal corrosion inhibitors is very scanty.[69,70] The reason for this observation would not be far from the insolubility of unmodified CPs in aqueous solutions. Thus, it would be helpful to extend the application of CPs as corrosion inhibitors by enhancing the solution insolubility. Potential strategies include the modification to polymer molecules with proper functional groups, the doping of water-soluble anions, and hydrophilic modification of polymer.

### 1.7.4 CP-Related Corrosion Mechanisms

In the past decade, the use of PANI as anti-corrosion coatings had been explored as the potential candidates to replace the chromium-containing materials, which have adverse health and environmental concerns. A polymer behaves as a barrier when it exists in the electronically and ionically insulating state. An important feature of the polymer coating in its conductive state is the ability to store large quantity of charge at the interface formed with a passive layer on a metal. This charge can be effectively used to oxidize a base metal to form a passive layer. Thus, the CP film was also capable of maintaining a stationary potential of the protected metal in the passive range.

There exist some differences between proposed mechanisms of corrosion protection of CPs, which largely depend on substrate metals, polymerization media, types of polymers, presence of topcoats, and synthesis methods. For example, some researchers attributed the improvement of the corrosion protection performance of the CPs to the formation of the underlying passive oxide film. The CP-based coating of PANI and PPy was also reported possessing self-healing properties, in which the passive oxide between the substrate metal and the CP could be spontaneously reformed at a flawed site by oxidative capability of the CP. Further, the anti-corrosion purpose can be greatly improved through the variation of the properties of CPs, such as size of the counter-ions and the inclusion of metallic oxide nanoparticles.

Several models of the possible corrosion protection mechanism are proposed according to the literatures.[47,66,71–76]

### 1.7.4.1 Ennobling Mechanism
One of the most extensively studied models is the so-called "ennobling mechanism" or "passivation mechanism". It is based on the assumption that the conductive polymer acts as an oxidizer and maintains metal in the passivity domain. This mechanism could induce the oxidation of the free metal surface at small defects in the passive layer.

### 1.7.4.2 Physical Barrier Mechanism
An alternative mechanism is known as a physical barrier mechanism. This protection mechanism is not unique to conducting polymers; that is, even nonconducting polymers can act as barriers. CPs, when coated properly, can separate the metal substrate and the corrosive environment. The absence of pinholes, cracks, and other coating defects guarantees effective protection. CPs can also be tuned to prevent diffusion of ions, water, and other aggressive species selectively. The coating blocks the diffusion of the ions and corrosion products, thus preventing the establishment of galvanic coupling between the local anodes and cathodes.

### 1.7.4.3 Ionic Perm-Selectivity Mechanism
The ionic perm-selectivity property of the polymer also play an important role in the performance of corrosion protection of ICPs. Since the

corrosion process includes the ionic transport from metal substrate to corrosion solution and opposite, therefore, polymers can behave like ionic membranes. It is important to consider the ion exchange property for the investigation of CPs. During the reduction of CPs (reaction 1.1), simultaneous release of anions takes place, from CP to solution (reaction 1.2), or incorporation of cations from solution to CP (reaction 1.3),

$$CP^{n+} + ne^- \rightarrow CP^0 \tag{1.1}$$

$$n/m \ A^{m-}_{(CP)} \rightarrow n/m \ A^{m-}_{(aq)} \tag{1.2}$$

$$n/e \ K^{l+}_{(aq)} \rightarrow n/l \ K^{l+}_{(CP)}. \tag{1.3}$$

The ion exchange behavior during charging and discharging processes of CPs has been widely investigated during past decade by electrochemical quartz crystal microbalance (EQCM) method.[77–80] The ion-exchange property of CPs is related to the polymerization conditions, types, and sizes of counter-ions incorporated into a CP matrix, its doping level, ions in the electrolyte solution, and pH. For example, PPy films doped with small anions such as:\ $Cl^-$, $NO^{3-}$, $ClO^{4-}$, or $SO_4^{2-}$ exhibit mainly anion exchange properties due to high mobility of these ions in a polymer matrix. The large-size organic anions, like polystyrene sulfonate (PSS), dodecyl sulfate (DoS), dodecylbenzenesulfonate (DBS), for doping would give the cationic permselective abilities to the CPs. From the viewpoint of corrosion protection, the PPy film exhibits a cation exchange property and may be a good barrier for attack of aggressive $Cl^-$ anions from an exposed solution.

### 1.7.4.4 Controlled Inhibitor Release Mechanism

According to this mechanism, the oxidized and hence doped CP-based coating deposited on a base metal substrate may release the anion dopant upon reduction, which is driven by a defect on the coating from coupling to the base metal. In the case of doped PANI, the anions are released not only through a reduction mechanism but also due to a simple elimination of acid-dopant if it is soluble in water.

### 1.7.4.5 Self-Healing Mechanism

The self-healing mechanism was based on the assumption that the CPs releases the doping anions, stored inside the conducting polymer matrix, when a coating defect appears. The doping anions diffuse into the defect site and decrease the corrosion rate. Besides the doped anions in the CPs matrix, many strategies have been proposed to fabrication the corrosion inhibitor containers by using CPs or CP composites. The passive oxide layer then re-forms between the metal and the conducting polymer brought about by the oxidative capability of the conducting polymer.[66]

## 1.7.5 Nanoparticiple-Encapsulated CPs

Application of CPs like PANI to corrosion protection of metals is, however, subject to some limitations. First, a charge stored in the polymer layer (used to oxidize base metal and to produce passive layer) can be irreversibly consumed during the system's redox reactions. Consequently, protective properties of the polymer coating may be lost with time. Also, porosity and anion exchange properties of CPs could be disadvantageous, particularly when it comes to pitting corrosion caused by small aggressive anions.

It would be helpful to enhance the protection properties of CP-based coatings by choosing proper synthesis conditions and by reasonable structure designing. In addition, the combination of CP and inorganic nanoparticles or 2D-nanosheets have shown the potential to overcome the above limitations. Many researchers have exploited the possibility of using composite materials based on CPs as anti-corrosive coatings. Composites combine the functional properties of diverse materials imparting improved properties for corrosion protection of metals and their alloys. A number of different metal, metalloid, metal oxide particles, or nanoparticles as well as carbon nanomaterials such as carbon nanotubes and graphene can be encapsulated into the matrix of a PANI to produce PANI-based composite materials.[3] Several critical reports have reviewed the recent progress of past and recent applications with CPs and polymer composites/nanocomposites.[30,57,72]

Researches are recommended to focus on developing new technologies that would widen the scope of application of CPs and CP/nanocomposites for corrosion protection. Further, new eco-environmental nanocomposited CP coatings with more effective protection are always welcome.

## 1.8 SUMMARY

*Corrosion* is a technical term used to describe the deterioration of a metal on reaction with its environment, causing serious economic loss and safety problems with human injuries and death. Techniques such as appropriate material design, material selection, cathodic and anodic protection, coatings, and modification of the environment through the use of corrosion inhibitors have been used to combat corrosion problem. Amongst these techniques, the use of coatings and corrosion inhibitors are the most popular and require the formation of a layer over a metallic substrate that offers a physical barrier against the access of aggressive species to the metallic interface.

Among the various polymers, conducting polymers, as a new corrosion-control methodology, have attracted more and more attention from researchers and related enterprises, owing to their excellent protective performance and environmental moderate features, especially under the requirements of sustainable development of the social environment. Conducting polymers as well as their nanocomposites can be used as pigment inhibitors in paints, layer or multilayers electrodeposited, and inhibitors in solutions. Corrosion-inhibitive dopants can be incorporated on the backbone of CPs, which can be released when CPs are reduced under particular conditions. Ionic-selective CPs could be synthesized by choosing proper electrolytes in an electrochemical synthesis strategy or chemical synthesis strategy.

## REFERENCES

1. McCafferty, E. 2010. *Introduction to Corrosion Science*. Springer, New York.
2. Hou, B., Li, X. & Ma, X. et al. 2017. The cost of corrosion in China. *NPJ Materials Degradation*, *1*, 1–10.
3. Umoren, S. A. & Solomon, M. M. 2019. Protective polymeric films for industrial substrates: A critical review on past and recent applications with conducting polymers and polymer composites/nanocomposites. *Progress in Material Science*, **104**, 380–450.
4. Fan, Y., Liu, X., Bai, K. & Wei, M. 2014. Research progress of anti-icing and deicing coatings. *New Chemical Materials*, **42**, 7–9.
5. Jones, A. D. 1995. *Principles and prevention of corrosion*, 2nd Edition, Pearson.
6. Antonijevic, M. M. & Petrovic, M. B. 2008. Copper corrosion inhibitors. a review. *International Journal of Electrochemical Science*, *3*, 1–28.
7. Twite, R. L. & Bierwagen, G. P. 1998. Review of alternatives to chromate for corrosion protection of aluminum aerospace alloys. *Progress in Organic Coatings*, *33*, 91–100.

8. Guo, X. & Facchetti, A. 2020. The journey of conducting polymers from discovery to application. *Nature Materials*, **19**, 922–928.

9. Alan, J. H., Alan, G. M. D. & Hideki, S. 2000. Nobel Prize in Chemistry. *The Nobel Foundation*.

10. Wang. X. X., Yu. G. F., Zhang. J., Yu. M., Seeram R. & Long. Y. Z. 2020,Conductive polymer ultrafine fibers via electrospinning: Preparation, physical properties and applications. Progress in Materials Science, **115**, 100704.

11. Bhandari, S. 2018. In *Polyaniline Blends, Composites, and Nanocomposites*, ed. Visakh, P. M., Della, P. C. & Falletta, E., 23–60. Elsevier Press.

12. Zhang, L., Du, W., Nautiyal, A., Liu, Z. & Zhang, X. 2018. Recent progress on nanostructured conducting polymers and composites: synthesis, application and future aspects. *Science China Materials*, **61**, 303–352.

13. Veerappan, G., Ramasamy, E. & Gowreeswari, B. 2019. In *Dye-Sensitized Solar Cells*, ed.Masoud, S. & Kenneth, K. S. L., 397–435. Academic Press.

14. Zhang, W., Yang, F. K., Pan, Z., Zhang, J. & Zhao, B. 2014. Bio-inspired dopamine functionalization of polypyrrole for improved adhesion and conductivity. *Macromolecular Rapid Communications*, **35**, 350–354.

15. Pang, A. L., Arsad, A. & Ahmadipour, M. 2021. Synthesis and factor affecting on the conductivity of polypyrrole: a short review. *Polymers for Advanced Technologies*, **32**, 1428–1454.

16. Kong, I. 2016. In *Design and Applications of Nanostructured Polymer Blends and Nanocomposite Systems*, ed. Thomas, S., Shanks, R. & Chandran, S., 125–154. William Andrew Publishing.

17. Xu, Y. *et al.* 2021. Poly(3,4-ethylenedioxythiophene) (PEDOT) as promising thermoelectric materials and devices. *Chemical Engineering Journal*, **404**, 126552.

18. Zeng, R. *et al.* 2021. CRISPR-Cas12a-driven MXene-PEDOT:PSS piezo-resistive wireless biosensor. *Nano Energy*, **82**, 105711.

19. Deng, W., Deng, L., Li, Z., Zhang, Y. & Chen, G. 2021. Synergistically boosting thermoelectric performance of PEDOT:PSS/SWCNT composites via the ion-exchange effect and promoting SWCNT dispersion by the ionic liquid. *ACS Applied Materials & Interfaces*, **13**, 12131–12140.

20. Wang, Y. 2009. Research progress on a novel conductive polymer–poly (3,4-ethylenedioxythiophene) (PEDOT). *Journal of Physics: Conference Series*, **152**, 012023.

21. Xue, S., Ma, Y., Miao, Y. & Li, W. 2020. Anti-corrosion performance of conductive copolymers of polyaniline/polythiophene on a stainless steel surface in acidic media. *International Journal of Nanoscience*, **19**, 012023.

22. Xue, S. & Liu, Q. 2016. Preparation of polythiophene/polypyrrole/TiO2 composite conductive polymers by solid-state method and its anti-corrosion properties for stainless steel. *Chinese Journal of Applied Chemistry*, **33**, 98–102.

23. Tüken, T., Yazıcı, B. & Erbil, M. E. H. M. E. T. 2005. Polypyrrole/poly-thiophene coating for copper protection. *Progress in Organic Coatings*, **53**, 38–45.
24. Tüken, T., Yazıcı, B. & Erbil, M. E. H. M. E. T. 2004. The use of poly-thiophene for mild steel protection. *Progress in Organic Coatings*, **51**, 205–212.
25. Kousik, G., Pitchumani, S. & Renganathan, N. G. 2001. Electrochemical characterization of polythiophene-coated steel. *Progress in Organic Coatings*, **43**, 286–291.
26. Kausar, A. 2019. Review on conducting polymer/nanodiamond nano-composites: Essences and functional performance. *Journal of Plastic Film & Sheeting*, **35**, 331–353.
27. Jiang, Y., Guo, X., Zhai, C. & Ding, W. 2002. Research Progress in Corrosion Prevention on Metal with Conductive Polymers. *Journal of Functional Polymer*, **15**, 473–479.
28. Gutierrez-Diaz, J. L., Uruchurtu-Chavarin, J., Guizado-Rodriguez, M. & Barba, V. 2016. Steel protection of two composite coatings: Polythiophene with ash or MCM-41 particles containing iron(III) nitrate as inhibitor in chloride media. *Progress in Organic Coatings*, **95**, 127–135.
29. De Leon, A. C., Pernites, R. B. & Advincula, R. C. 2012. Superhydrophobic colloidally textured polythiophene film as superior anti-corrosion coating. ACS Appl Mater Interfaces, **4**, 3169–3176.
30. Poddar, A. K., Patel, S. S. & Patel, H. D. 2021. Synthesis, characterization and applications of conductive polymers: a brief review. *Polymers for Advanced Technologies*, **32**, 4616–4641.
31. Jia, X., Ge, Y., Shao, L., Wang, C. & Wallace, G. G. 2019. Tunable con-ducting Polymers: Toward Sustainable and Versatile Batteries. *ACS Sustainable Chemistry & Engineering*, **7**, 14321–14340.
32. Lewandowski, A., Swiderska-Mocek, A. & Waliszewski, L. 2013. Li+ conducting polymer electrolyte based on ionic liquid for lithium and lithium-ion batteries. *Electrochimica Acta*, **92**, 404–411.
33. Inzelt, G. 2017. Recent advances in the field of conducting polymers. *Journal of Solid State Electrochemistry*, **21**, 1965–1975.
34. Lei, Y. *et al.* 2020. Synthesis of porous n-rich carbon/mxene from mxene@polypyrrole hybrid nanosheets as oxygen reduction reaction electrocatalysts. *Journal of the Electrochemical Society*, **167**, 116530
35. Daş, E. & Yurtcan, A. B. 2016. Effect of carbon ratio in the polypyrrole/carbon composite catalyst support on PEM fuel cell performance. *International Journal of Hydrogen Energy*, **41**, 13171–13179.
36. Wu, G., More, K. L., Johnston, C. M. & Zelenay, P. 2011. High-performance electrocatalysts for oxygen reduction derived from poly-aniline, iron, and cobalt. *Science*, **332**, 443–447.
37. Ramírez-Pérez, A. C., Quílez-Bermejo, J., Sieben, J. M., Morallón, E. & Cazorla-Amorós, D. 2018. Effect of nitrogen-functional groups on the orr

activity of activated carbon fiber-polypyrrole-based electrodes. *Electrocatalysis*, **9**, 697–705.

38. Tan, N. *et al.* 2022. Fabricating Pt/CeO₂/N-C ternary ORR electrocatalysts with extremely low platinum content and excellent performance. *Journal of Materials Science*, **57**, 538–552.

39. Lei, Y. *et al.* 2021. Fabrication of porous n-rich carbon electrocatalysts frompyrolysis of for e. *Journal of the Electrochemical Society*. **168**, 044516.

40. Zhang, Y. *et al.* 2020. In situ chemo-polymerized polypyrrole-coated filter paper for high-efficient solar vapor generation. *International Journal of Energy Research*, **44**, 1191–1204.

41. Yan, J. *et al.* 2010. Preparation of a graphene nanosheet/polyaniline composite with high specific capacitance. *Carbon*, **48**, 487–493.

42. Ren, L. *et al.* 2015. Three-Dimensional Tubular MoS2/PANI Hybrid Electrode for High Rate Performance Supercapacitor. *ACS Appl Mater Interfaces*, **7**, 28294–28302.

43. Gao, Z. *et al.* 2013. Electrochemical synthesis of layer-by-layer reduced graphene oxide sheets/polyaniline nanofibers composite and its electrochemical performance. *Electrochimica Acta*, **91**, 185–194.

44. Le, T. A., Tran, N. Q., Hong, Y. & Lee, H. 2019. Intertwined Titanium Carbide MXene within a 3 D Tangled Polypyrrole Nanowires Matrix for Enhanced Supercapacitor Performances,25. *Chemistry*, 1037–1043.

45. Qin, L. *et al.* 2018. High-Performance Ultrathin Flexible Solid-State Supercapacitors Based on Solution Processable Mo1.33C MXene and PEDOT:PSS. *Advanced Functional Materials*, **28**, 1703808.

46. Karimullah, A. S., Cumming, D. R. S., Riehle, M. & Gadegaard, N. 2013. Development of a conducting polymer cell impedance sensor. *Sensors and Actuators B: Chemical*, **176**, 667–674.

47. Pillay, V. *et al.* 2014. A review of integrating electroactive polymers as responsive systems for specialized drug delivery applications. *Journal Biomedical Materials Research A*, **102**, 2039–2054.

48. Ates, M. 2013. A review study of (bio)sensor systems based on conducting polymers. *Materials Science & Engineering C-Materials for Biological Applications*, **33**, 1853–1859.

49. Bhadra, J., Al-Thani, N. J., Madi, N. K. & Al-Maadeed, M. A. 2013. Preparation and characterization of chemically synthesized polyaniline–polystyrene blends as a carbon dioxide gas sensor. *Synthetic Metals*, **181**, 27–36.

50. Le, T. H., Kim, Y. & Yoon, H. 2017. Electrical and electrochemical properties of conducting polymers. *Polymers*, **9**, 150.

51. Tallman, D. E., Spinks, G., Dominis, A. & Wallace, G. G. 2001. Electroactive conducting polymers for corrosion control. *Journal of Solid State Electrochemistry*, **6**, 73–84.

52. Cascalheira, A. C., Aeiyach, S., Lacaze, P. C. & Abrantes, L. M. 2003. Electrochemical synthesis and redox behaviour of polypyrrole coatings on copper in salicylate aqueous solution. *Electrochimica Acta*, **48**, 2523–2529.

53. Bierwagen, G. P. 1998. *Corrosion and Its Control by Coatings*. 1–8.

54. Jafari, Y., Ghoreishi, S. M. & Shabani-Nooshabadi, M. 2016. Polyaniline/ Graphene nanocomposite coatings on copper: Electropolymerization, characterization, and evaluation of corrosion protection performance. *Synthetic Metals*, **217**, 220–230.

55. Ma, Y. *et al.* 2020. Enhanced corrosion inhibition of aniline derivatives electropolymerized coatings on copper: Preparation, characterization and mechanism modeling. *Applied Surface Science*, **514**, 146086.

56. Liu, H. *et al.* 2021. Anti-corrosive mechanism of poly (N-ethylaniline)/ sodium silicate electrochemical composites for copper: Correlated experimental and in-silico studies. *Journal of Materials Science & Technology*, **72**, 202–216.

57. Zadeh, M. K., Yeganeh, M., Shoushtari, M. T. & Esmaeilkhanian, A. 2021. Corrosion performance of polypyrrole-coated metals: A review of perspectives and recent advances. *Synthetic Metals*, **274**, 116723.

58. Mondal, S., Das, S. & Nandi, A. K. 2020. A review on recent advances in polymer and peptide hydrogels. *Soft Matter*, **16**, 1404–1454.

59. Kiari, M., Berenguer, R., Montilla, F. & Morallon, E. 2020. Preparation and Characterization of Montmorillonite/PEDOT-PSS and Diatomite/ PEDOT-PSS Hybrid Materials. Study of Electrochemical Properties in Acid Medium. *Journal of Composites Science*.4, 51.

60. Menkuer, M. & Ozkazanc, H. 2019. Electrodeposition of polypyrrole on copper surfaces in OXA-DBSA mix electrolyte and their corrosion behavior. *Progress in Organic Coatings*, **130**, 149–157.

61. DeBerry, D. W. 1985. Modification of the Electrochemical and Corrosion Behavior of Stainless Steels with an Electroactive Coating. *Journal of The Electrochemical Society*, **132** 1022–1026.

62. Qiu, S., Zhao, H. & Wang, L. 2018. Facile preparation of soluble poly(2-aminothiazole)-based composite coating for enhanced corrosion protection in 3.5% NaCl solution. *Surface Topography: Metrology and Properties*, **6**, 034007.

63. Deshpande, P. P., Bhopale, A. A., Mooss, V. A. & Athawale, A. A. 2016. Conducting polyaniline/nano-zinc phosphate composite as a pigment for corrosion protection of low-carbon steel. *Chemical Papers*, **71**, 189–197.

64. Lei, Y., Sheng, N., Hyono, A., Ueda, M. & Ohtsuka, T. 2013. Electrochemical synthesis of polypyrrole films on copper from phytic solution for corrosion protection. *Corrosion Science*, **76**, 302–309.

65. Kartsonakis, I. A., Balaskas, A. C., Koumoulos, E. P., Charitidis, C. A. & Kordas, G. C. 2012. Incorporation of ceramic nanocontainers into epoxy coatings for the corrosion protection of hot dip galvanized steel. *Corrosion Science*, **57**, 30–41.

66. De Leon, A. & Advincula, R. C. 2015. In *Intelligent Coatings for Corrosion Control*, ed. Atul T., James R. & Lloyd H. H., 409–430. Butterworth-Heinemann Press.

67. Umoren, S. A. & Solomon, M. M. 2014. Recent developments on the use of polymers as corrosion inhibitors - a review. *The Open Materials Science Journal*, **8**, 39–54.
68. Fateh, A., Aliofkhazraei, M. & Rezvanian, A. R. 2020. Review of corrosive environments for copper and its corrosion inhibitors. *Arabian Journal of Chemistry*, **13**, 481–544.
69. Jeyaprabha, C., Sathiyanarayanan, S. & Venkatachari, G. 2006. Effect of cerium ions on corrosion inhibition of PANI for iron in 0.5M $H_2SO_4$. *Applied Surface Science*, **253**, 432–438.
70. Sathiyanarayanan, S., Balakrishnan, K., Dhawan, S. K. & Trivedi, D. C. 1994. Prevention of corrosion of iron in acidic media using poly (o-methoxy-aniline). *Electrochimica Acta*, **39**, 831–837.
71. Deshpande, P. P., Jadhav, N. G., Gelling, V. J. & Sazou, D. 2014. Conducting polymers for corrosion protection: a review. *Journal of Coatings Technology and Research*, **11**, 473–494.
72. Samadzadeh, M., Boura, S. H., Peikari, M., Kasiriha, S. M. & Ashrafi, A. 2010. A review on self-healing coatings based on micro/nanocapsules. *Progress in Organic Coatings*, **68**, 159–164.
73. Kim, H., Lee, H., Lim, H. R., Cho, H. B. & Choa, Y. H. 2019. Electrically conductive and anti-corrosive coating on copper foil assisted by polymer-nanocomposites embedded with graphene. *Applied Surface Science*, **476**, 123–127.
74. Hendy, G. M. & Breslin, C. B. 2019. The incorporation and controlled release of dopamine from a sulfonated β–cyclodextrin–doped conducting polymer. *Journal of Polymer Research*, **26**, 1–8.
75. Abdullatef, O. A. & Farid, R. M. 2018. Electropolymerization of mefe-namic acid on copper and copper based alloy as a new strategy to control the release of copper ions from copper containing devices. *Russian Journal of Applied Chemistry*, **90**, 1866–1875.
76. Rupali Gangopadhyay, A. D. 2000. Conducting Polymer Nanocomposites A Brief Overview. *Chem. Mater*, **12**, 608–622.
77. Weidlich, C., Mangold, K. M. & Jüttner, K. 2005. EQCM study of the ion exchange behaviour of polypyrrole with different counterions in different electrolytes. *Electrochimica Acta*, **50**, 1547–1552.
78. Debiemme-Chouvy, C., Cachet, H. & Deslouis, C. 2006. Investigation by EQCM of the electrosynthesis and the properties of polypyrrole films doped with sulphate ions and/or a Keggin-type heteropolyanion, SiMo12O404–. *Electrochimica Acta*, **51**, 3622–3631.
79. Koehler, S., Ueda, M., Efimov, I. & Bund, A. 2007. An EQCM study of the deposition and doping/dedoping behavior of polypyrrole from phos-phoric acid solutions. *Electrochimica Acta*, **52**, 3040–3046.
80. Lei, Y. H., Sheng, N., Hyono, A. & Ueda, M. 2014. Effect of benzotriazole (BTA) addition on Polypyrrole film formationon copper and its corrosion protection. *Progress in Organic Coatings*, **77**, 339–346.

# CHAPTER 2

# Corrosion Protection of Carbon Steel by Conducting Polypyrroles

Toshiaki Ohtsuka

## CONTENTS

## 2.1 INTRODUCTION

A conducting polymer (CP) has been found since 1980 by Shirakawa et al.[1] and its application has been extended to a wide range of industrial products. Typical CPs are PANI, PPy, and PTh, as shown in Figure 2.1. The application of corrosion protection of metallic materials

DOI: 10.1201/9781003376194-2

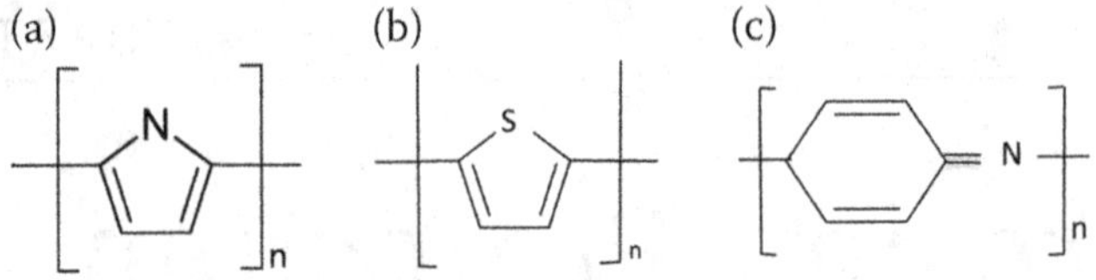

(a)　　　　　　　(b)　　　　　　(c)

FIGURE 2.1 Typical conducting polymers; (a) polypyrrole (PPy), (b) poly-thiophen (PThio), and (c) polyanniline (PANI).

**Source: created by the authors.**

by CPs was reported by DeBerry et al.[2] and many reports have been published.[3-6] Wessling proposed a model of corrosion protection of metals in which the oxidative property of the CP covering the metals induces a potential shift in the positive direction and passivation of the metals underneath the CP.[7] Under the passive state thin oxide film several nm thick covers the substrate metal and the corrosion rate is kept at an extremely small degree. The application of the CP coating to the corrosion protection of steels was reviewed by Tallman et al.[8]

The duration in which the CP keeps its oxidative property is assumed to depend on the doped anions in the CP. When the anions doped possess a relatively large mobility, the anions are easily removed (or undoped) from the CP, inducing a loss of the oxidative property. The loss of the oxidative property is combined with the instability of a passive oxide layer on metal, causing a transition from the passive state to the active state in a relatively short period. The protection in the passive state is thus not assumed to be kept for a long period.

The CP can work as a physical barrier against the penetration of water, oxygen, and aggressive anions such as chloride ions, too. In order to enhance the barrier effect, the inert particles have been included in $SiO_2$, $Fe_2O_3$, graphene, etc. and as a consequence of the enhancement of the barrier effect, the protection property of CP has been reported to be much improved.

In this section, we pay attention to the corrosion protection by the PPy layer and the corrosion protection of steels by the PPy layer is discussed from a viewpoint of a redox property and a barrier effect of the PPy.

## 2.2 OXIDATIVE POLYMERIZATION OF POLYPYRROLE

Electrochemical polymerization by oxidation of pyrrole (Py) monomers have been reviewed by many authors and here we briefly describe the

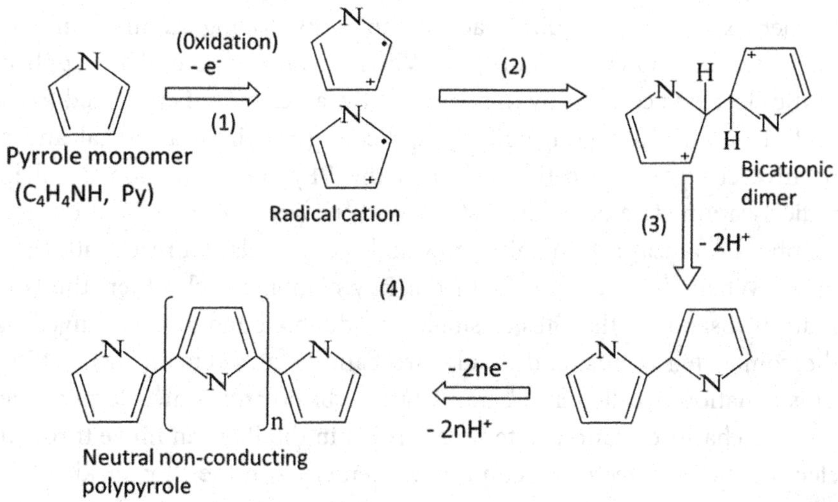

FIGURE 2.2 Electropolymerization process of PPy.

Source: created by the authors.

process of PPy. When an electrode is anodically polarized in an electrolyte solution containing Py monomers, the black polymer film can be formed on the electrode. The polymerization procedure is performed without any difficulty, except for careful treatment of the electrolyte in which oxidation of the Py monomer by remaining air should be avoided. The electrolyte should be thus deoxygenated by inert gas bubbling before injection of a Py monomer.

Figure 2.2 illustrates a model of the process for oxidation polymerization of PPy proposed by Genies and Biden.[9] The Py monomer dissolved in the electrolyte donates electrons into the electrode, forming a radical-cation pair (step(1)). The radicals in Py monomers are reacted with each other and two protons are simultaneously removed from the reacted Py pair (step (2)), forming a dimer of Py (step (3)). After the formation of the radical-cation pair and the reaction between the radicals are repeated, the black PPy film is deposited on the electrode (step (4)).

## 2.3 OXIDATION AND REDUCTION OF POLYPYRROLE AND ITS CONDUCTING PROPERTIES

The neutral PPy formed with a conjugated chain does not exhibit any electronic conductivity. To add the conductivity into the neutral PPy, further oxidation is required, accompanied by doping anions from the electrolyte, as shown in Figure 2.3. When the anodic potential is applied to the electrode covered by the neutral PPy, an electron is removed from π electrons in the conjugated bond, yielding a pair of a radical and a positive charge (or a cation state) in the PPy. This situation is called radical-cation state or polaron state. With progress of the oxidation, the number of radical-cation pairs grow and the radicals can react with each other. When the two radicals in the PPy combine each other, the two radicals disappear, the sites of single and double bond are exchanged in the conjugated bonds, and finally two cation states remain in the PPy. This situation is called a bi-cation state or bi-polaron state. Because the positive charge or cation state thus formed in the PPy can move through electron clouds, electronic conductivity emerges in the PPy backbone.

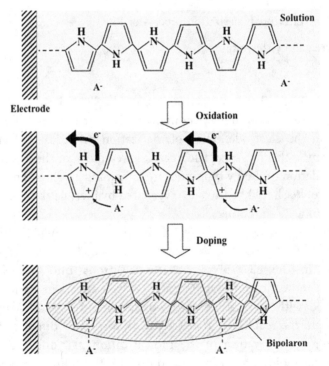

FIGURE 2.3   Electrochemical oxidation of neutral non-conducting PPy. During the oxidation, electron transfer from PPy to substrate steel and doping of anions from electrolyte solution to PPy simultaneously occur. In the progress of oxidation, two radicals are combined with each other and disappear, forming a bi-cation or bi-polaron state.

Source: *International Journal of Corrosion* (Open access).

With the removal of electrons from the PPy backbone, injection of anions from an electrolyte solution simultaneously occurs to maintain neutrality of the PPy layer; i.e., when the neutral state of PPy backbone changed to the oxidative state, removal of electrons and doping of anions simultaneously take place.

$$PPy + (n/x)A^{x-} \rightleftharpoons PPy^{n+} \cdot (n/x)A^{x-} + ne^- \tag{2.1}$$

It has been suggested that one positive charge site (or one cation state) was inserted in three or four Py units at the maximum. The maximum amount of the anions doped are determined by the corresponding positive charge. When the more positive charge is added, the PPy changes to the over-oxidation state and loses the conductivity.

## 2.4 PROTECTION MECHANISMS OF CONDUCTING POLYMERS

Two mechanisms have been proposed for the corrosion protection of the conducting PPy. One is anodic protection and the other is a barrier effect. The anodic protection is caused by a strong oxidant property of the PPy, which shifts the potential of the substrate steel to the positive direction, bringing about the passive state of the steel, as shown in Figure 2.4.[2,7,8] In Figure 2.4, a schematic model of anodic current density (CD ($i_a$)) of the steel and cathodic CD ($i_c$) of the surrounding compounds of $O_2$, $H^+$, and $H_2O$ against potential ($E$). When the steel is covered by the conducting PPy, oxidation and reduction CDs of the PPy are added on the original $E$-$i$ relation of the steel[10].

$$PPy^{n+} \cdot (n/x)A^{x-} + \Delta ne^- \rightleftharpoons PPy^{(n-\Delta n)+} \cdot (n - \Delta n/x)A^{x-} + (\Delta n/x)A^{x-}_{aq} \tag{2.2}$$

The redox potential of the PPy depends on oxidation state of the PPy, $PPy^{n+} \cdot (n/x)A^{x-}$, increasing with the increase of oxidation number n. When the redox exchange CD of the oxidative PPy is much higher than the CD of the passive steel, the potential of the steel covered by the PPy is determined by the redox potential of the oxidative PPy. The corrosion potential of the bare steel is usually located in the active dissolution region and is shifted to the passive potential region when the steel is covered by the oxidative PPy. As described in Figure 2.4, the current

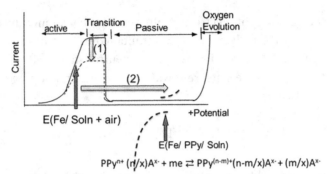

$$PPy^{n+} (n/x)A^{x-} + me \rightleftarrows PPy^{(n-m)+}(n-m/x)A^{x-} + (m/x)A^{x-}$$

Role of oxidative-conductive polymer for corrosion prevention
(1) Suppression of active dissolution = barrier effect
(2) Potential shift by oxidative polymer = anodic protection

FIGURE 2.4 Potential (E) current (i) relation of steels covered by oxidative conducting PPy. A barrier effect of PPy suppresses active dissolution of the steel and an oxidative property of PPy shifts the potential into passive state.

Source: *International Journal of Corrosion* (Open access).

density in the active region should also be decreased by a barrier effect of the PPy layer. In the passive state to which the steel covered by the PPy attains, corrosion current density may remain lower than 1 mA cm$^{-2}$ or the corrosion rate is lower than 0.02 mm y$^{-1}$.

The polymer coating can work as a chemical and physical barrier against the penetration of water, oxygen gas, and aggressive anions, inducing a decrease of the corrosion rate. The barrier effect also affects the anodic dissolution of the steel substrate:

$$Fe \rightarrow Fe^{2+} + 2e \tag{2.3}$$

decreasing the anodic peak in the active dissolution region, as shown in Figure 2.4. The decrease of the active peak facilitates the potential shift from the active to the passive region by the oxidative PPy layer. When the PPy layer plays the role of an enough barrier, the corrosion of the steel is much restricted, the potential of which is even in active region.[11] In order to enhance the barrier effect, attempts have been made to embed solid particles of metal oxides, graphite tube, etc. in the PPy layer.[12–18] When the particles occupy a large ratio, the role of the barrier is much increased and, however, the conductivity and the redox property are decreased more with the increasing amount of the solid particles. In the large amount of particles, the potential shift to the passive region may not be able to occur and, however, the corrosion may decline by the barrier effect of the PPy layer including the inert particles.[9]

## 2.5 DESIGN OF PPY COATING FOR CORROSION PROTECTION

### 2.5.1 Role of Doping Anions and Solid Powders

The conducting PPy includes the doped anions according to equation of 2.1. The stability of the passive state of steels or the passive oxide on steels depends on the anions. Phosphate, molybdate, tungstate, etc. are classified to a passivating inhibitor that facilitates the passive state in the presence of oxidants and stabilizes the passive state. When such anions are doped to the PPy layer that contacts the steel substrate, the passive oxide film on the steel is stabilized and the period of the passive state is prolonged. The doped anions are therefore very important to keep the passive state for a long period for inhibition of the steel corrosion.

### 2.5.2 Ionic Permselectivity of the Polypyrrole Layer

The release of doped anions from the PPy layer and exchange between the doped anions and anions in electrolytes are affected by diffusion or migration rate of anions in the PPy layer and the rate depends on the size and mass of the anions. When usual mineral acid ions are doped, the exchange reaction of anions occurs between the PPy layer and electrolyte (Figure 2.5(a)).

$$PPy^{n+} \cdot (n/x) A^{x-} + (n'/y) B^{y-} \rightleftarrows PPy^{n+}$$

$$\cdot [(n - \Delta n)/x] A^{x-} \cdot (\Delta n'/y) A^{y-} + [(n' - \Delta n')/y] B^{y-} \qquad (2.4)$$

where $A^{x-}$ and $B^{y-}$ are anions in the PPy layer and electrolyte, respectively, and $n/x = n'/y$. With the exchange reaction, the anions in the electrolyte can penetrate through the PPy layer. The PPy layer doped with a small anion works as an anionic permselective membrane, in which anions are preferentially transported by a gradient of electrochemical potential. If the anions are aggressive anions like chloride and bromide ions, the anions penetrating into the steel substrate induce breakdown of the passive oxide film. The breakdown of the passive film is followed by a large damage of localized corrosion of pitting and crevice corrosion.

When the size of the anions is large, the migration rate is so small that the movement of the anions is restrained. Instead of anion transport,

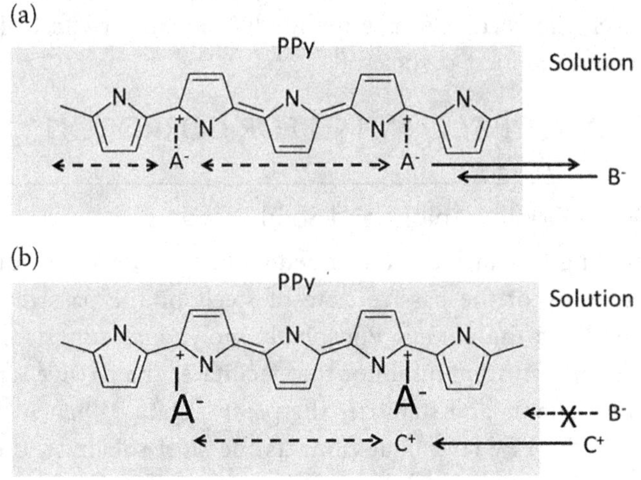

(a)

(b)

FIGURE 2.5 Ionic permselectivity of PPy film. (a) PPy film with anionic permselectivity, in which small-size anions are doped in PPy and (b) PPy film with cationic permselectivity, in which large-size anions are doped in PPy film.

Source: *International Journal of Corrosion* (Open access).

cations penetrates through the PPy layer and thus the PPy layer works as a cationic permselective membrane in which cations are preferentially transported (Figure 2.5(b)). The PPy layer doped with anions of a large size thus inhibit penetration of aggressive anions from the electrolyte and protect the steel against the localized corrosion.

The ionic permselective property is examined from the mass change with oxidation or reduction of the PPy layer. With reduction of the PPy layer, the release of anions of small size from the PPy layer to the electrolyte is caused (reaction (2.4)). For a PPy layer doped with anions of a large size, the cations are inserted from the electrolyte into the PPy layer with a reduction of the PPy layer:

$$PPy^{n+} \cdot (n/x) A^{x-} + (c/z) C^{z+} + \Delta ne- \rightleftarrows PPy^{(n+\Delta n)+}$$
$$\cdot nA^{x-} \cdot (\Delta c/z) C^{z+} + [(c - \Delta c)/z] C^{z+} \tag{2.5}$$

where $C^{z+}$ is cation in the electrolyte. When the mass gain is observed with the reduction, the uptake of cations into the PPy layer occurs and the cations are mobile in the layer. Conversely, when the mass gain observed with the oxidation, the anions are doped into the PPy layer and mobile in the layer. Electrochemical quartz crystal microbalance (EQCM) has been available for the measurement of a mass change of the PPy layer on the electrode.[19]

## 2.6 EVALUATION OF CORROSION RATE OF STEEL COVERED BY POLYPYRROLES

Corrosion protection of the steels covered by the PPy layer was examined by many methods. Electrochemical techniques of the Tafel line extrapolation, linear polarization resistance, and AC impedance or electrochemical impedance spectroscopy (EIS) are available for the evaluation and, however, we must take notice of the application of electrochemistry.[20]

When one measures the passive steel covered with a conductive and redox PPy layer by electrochemistry, an electrochemical response from the redox PPy preferentially emerges rather than that from corrosion of the steel substrate. Because of the response current originates in the redox reaction of the PPy layer, the exchange current estimated from the above electrochemical techniques is not the corrosion current, but an exchange current of the redox reaction of the PPy layer described in (Eq. 2.2).[21] In this case, the corrosion rate may be estimated from the weight loss of the steel or the amount of released Fe ions from the steel during an immersion test in a solution containing aggressive species.

It is possible that the corrosion current of the steel is detected from the electrochemical techniques only when the PPy layer loses the conductive and redox properties. In the case that the corrosion environment is a neural or alkaline pH aqueous solution, the brown rust of iron can be seen immediately with the corrosion and thus the progress of corrosion can be judged from the appearance of the rust.

## 2.7 PERFORMANCE OF CORROSION PROTECTION OF CARBON STEEL BY BILAYERED POLYPYRROLE

The anodic protection greatly depends on the stability of passivity and passive oxides on the steel. For example, phosphate and molybdate ions work as a species stabilizing the passive oxide on the steel. For the stability, attack of aggressive anions such as chloride and bromide also play an important role and they initiate breakdown of the passive oxide followed by local corrosion of pitting corrosion and stress cracking corrosion (SCC). If penetration of the anions can be inhibited by a coating layer, the local corrosion will be stopped.

Deslouis et al. anodically prepared a PPy layer on steel in an oxalate solution containing Py and reported that the PPy layer can protect the steel from corrosion in a sodium chloride solution, keeping the passive state of the steel for long time period.[22–24] They assumed that the ferric oxalate layer that was formed underneath the PPy film during anodic oxidation before PPy polymerization worked as a passivation film against corrosion. They further reported that an outer layer of PPy doped with a dodecylsulfate ion, $C_{12}H_{25}OSO_3^-$ (DoS), was prepared on the inner PPy layer doped with oxalate ions. The bilayer PPy coating was effective to the corrosion protection and maintained the passivation state for longer than 500 h, in which no corrosion products of iron rust were observed.

A DoS ion functions as a surfactant and forms a micelle in an aqueous solution at a higher than critical concentration. Py monomers added in an aqueous solution are incorporated in the micelle of a DoS ion. When an anodic potential is applied to an electrode, the micelles are collapsed on the electrode and the polymerization of Py monomers starts to form a PPy layer on the electrode. Because a DoS ion has a relatively large mass and volume, its mobility is small in the PPy layer and thus it works as an immobile dopant. The PPy layer doped with a DoS ion (PPy-DoS) is, therefore, considered a membrane with negatively charged fixed sites and exhibits cationic perm-selectivity.[14] The outer layer of PPy-DoS can exclude the insertion of aggressive anions such as chloride ions. On the PPy layer with a redox property and cationic perm-selectivity, cation uptake and release occur with reduction and oxidation, respectively.

Kowalski et al. reported a bilayered PPy coating for corrosion protection of carbon steels.[25–27] They prepared the PPy layers doped with phosphate, $PO_4^{3-}$, and phospho-molybdate ion, $PMo_{12}O_{40}^{3-}$ (PMo), as an inner layer and the PPy layer doped with a DoS ion as an outer layer. They evaluated the corrosion protection from the period in which the passivity was maintained in a sodium chloride (3.5% NaCl) solution. Both phosphate and molybdate ions doped in the inner layer works as passivating inhibitors, which facilitates passivity of the steel and stabilizes the passive oxide on the steel. The outer layer doped with a DoS excludes the insertion of chloride ions. As shown in Figure 2.6, the passivity was maintained in 3.5% sodium chloride solution for 170 h for which the dissolution of Fe and the rust formation were not observed. After 170 h of immersion, the corrosion potential started to gradually change in the negative direction, reaching the potential in the active region. The brown rust appeared immediately after the potential reached the active region.

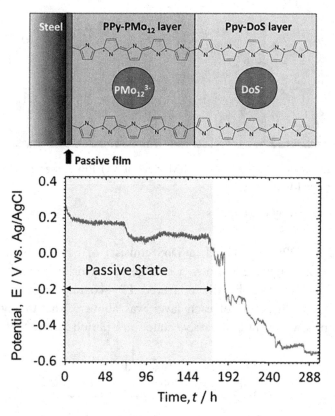

FIGURE 2.6 Model of bilayered PPy film prepared by Kowalski et al. and transient of open circuit potential of steel covered by the bilayered PPy in 3.5% NaCl solution. The inner and outer layers was prepared by constant CD of 1.0 mA cm-2 for 1,000 s in solution of 0.1 M Py monomer, 0.2 M H3PO4, and 5 mM H3PMo12O40, and in a solution of 0.1 M Py monomer 25 mM sodium dodecylsulfate (NaDoS), respectively. The thickness of each layer is about 5 mm.

**Source:** *International Journal of Corrosion* (**Open access**).

Sasaki et al. prepared on a carbon steel a bilayered PPy coating that consisted of the inner PPy layer doped with PMo ion and the outer PPy doped with DoS.[28] The bilayered PPy coating was first formed by constant current oxidation at 1.0 mA cm$^{-2}$ for 1,000 s in a solution of 15 mM PMo containing 0.1 M Py monomer and the outer layer at at 1.0 mA cm$^{-2}$ for 1000 s in solution of 5 mM sodium DoS containing 0.1 M Py monomer. Figure 2.7 shows microscopic views observed by scanning electron microscopy (SEM), in which the PPy doped with PMo is seen to consist of fine particles with a diameter less than 100 nm,

FIGURE 2.7   Microscopic surface SEM view of (A) PPy doped with PMo and (B) PPy doped with DoS.

**Source: Created by the authors.**

whereas the PPy doped with a DoS consists of particles with several hundred nm. Figure 2.8 shows a change in corrosion potential of the carbon steel covered by the bilayered PPy coating in a 3.5% NaCl solution. The thickness of each layer was about 5 μm. In Figure 2.8, the potential was kept at a passive state for a period longer than 1,000 h

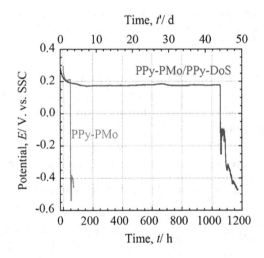

FIGURE 2.8   Transient of open circuit potential of steel covered by the bilayered PPy in a 3.5% NaCl solution. The inner (PPy-PMo) and outer (PPy-DoS) layers were prepared by a constant CD of 1.0 mA cm$^2$ for 1,000 s in solution of 0.1 M Py, monomer and 5 mM H3PMo12O40, and in a solution of 0.1 M Py, monomer 25 mM sodium dodecylsulfate (NaDoS), respectively. The thickness of each layer is about 5 μm.

**Source: Created by the authors.**

(or 1.5 months). In the figure, the result for the steel covered by a single PPy layer doped with PMo is shown as a comparison. The period of a passive state is only about 50 h without the outer PPy layer with a DoS.

When enough conductivity kept through the PPy layer, the oxidant property of the layer becomes proportional to its thickness. Because the coating with the larger thickness plays a role of the larger barrier against an ion and molecule transfer, too, the thicker coating can possess the higher protection capability. When one prepared the bilayer PPy coating five times thicker than that in Figure. 2.8, the steel covered was kept at a passive state for period even longer than 3,000 h (four months).[29]

Because the PPy layer possibly inhibits the penetration of chloride ions to the steel, the steel covered by the PPy layer is assumed to be protected by pitting corrosion induced by chloride ions. Figure 2.9 shows a potentiodynamic i-E curve measured for bare steel and PPy-coated steels at a sweep rate of 20 mV min$^{-1}$ in a 3.5% NaCl solution.[29] Current density (CD) of the bare steel sharply increased from a low potential at –0.3 V (vs. Ag/AgCl/ saturated KCl), revealing the initiation of pitting corrosion at the potential. For the steel covered by a PPy layer, an increase of CD was started at the potential much higher than that for

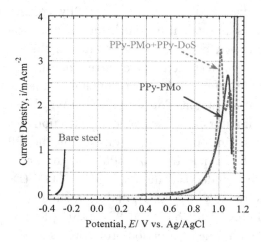

FIGURE 2.9   Potentio-kinetic i-E curve of bare steel and the steel covered by a PPy coating at a sweep rate of 20 mV min$^{-1}$ in a 3.5% NaCl solution. PPy-PMo is a single PPy coating doped PMo and PPy-PMo+PPy-DoS is a bilayered PPy coating consisting of an inner PPy layer doped with PMo and an outer PPy layer doped with DoS. The preparation method is the same as that in Figure 2.8.

**Source: created by the authors.**

the bare steel. The steel covered by a PPy layer doped with PMo exhibited an increase of CD at 0.8 V, revealing a maximum at 1.07 V and then decreased. The CD maximum is assumed to be overoxidation in which hydroxyl radicals formed by water oxidation attacks the site of a Py ring to form C-OH and C=O. After the overoxidation, PPy loses conductivity and redox properties irreversibly.[30-34] The following sharp increase of CD may be due to an oxygen evolution reaction or propagating breakdown of the layer. The steel covered by the bilayered coating of the inner PPy doped with PMo and the outert PPy doped with DoS revealed two maximums of CD in which the first maximum at the lower potential is assumed to correspond to an overoxidation peak of the inner PPy doped with PMo and the second maximum in the higher potential to an overoxidation of the outer PPy doped with a DoS. Nevertheless, the pitting corrosion is not seen on the steel covered by the PPy layer until the overoxidation of PPy layer and oxygen evolution reaction is started and the PPy layer effectively inhibits pitting corrosion initiated by a chloride ion attack.

## 2.8 SUMMARY

Protection of steel by a conducting polypyrrole (PPy) coating against corrosion was discussed. The conducting PPy coating with a redox property works as an oxidant by which the steel underneath the PPy is shifted from the active to the passive state. In the passive state, the corrosion rate of the steel decreases to current densities, CDs, lower than $1$ mA cm$^{-2}$ or $0.02$ mm y$^{-1}$. In order to maintain the passive state for a long time period, the anions doped to the PPy may be required to work as passivating inhibitors. As an example of protection of steel, a bilayered PPy coating was introduced that consists of the PPy doped with phospho-molybdate ion, $PMo_{12}O_{40}^{3-}$ (PMo), as an inner layer and of the PPy doped with a dodecylsulfate ion, $C_{12}H_{25}OSO_3^-$ (DoS) as an outer layer. The bilayered PPy coating $10$ μm thick maintained the passive state for about $1,000$ h and the bilayered PPy coating $50$ μm thick for even longer than $3,000$ h. The long-period protection is due to the inner layer doped with PMo that stabilizes the passive oxide on the steel and to the outer layer doped with a DoS that inhibits penetration of chloride ions from electrolyte to the steel surface. In order to enhance the barrier effect of the PPy layer, it was also attempted to embed the inert solid particles in the PPy layer.

# REFERENCES

1. Shirakawa, H., Louis, E. J. & Gau, S. C. 1977. Electrical conductivity in doped polyacetylene. *Physical review letters*, **39**, 1098.
2. DeBerry, D. W. 1985. Modification of the electrochemical and corrosion behavior of stainless steels with an electroactive coating. *Journal of the Electrochemical Society*, **132**, 1022.
3. Konno, Y., Farag, A. A. & Tsuji, E. 2016. Formation of porous anodic films on carbon steels and their application to corrosion protection composite coatings formed with polypyrrole. *Journal of the Electrochemical Society*, **163**, C386–C393.
4. Garcia-Cabezon, C., Garcia-Hernandez, C. & Rodriguez-Mendez, M. L. 2020. A new strategy for corrosion protection of porous stainless steel using polypyrrole films. *Journal of Materials Science & Technology*, **37**, 85–95.
5. Souto, L. F. C. & Soares, B. G. 2020. Polyaniline/carbon nanotube hybrids modified with ionic liquids as anticorrosive additive in epoxy coatings. *Progress in Organic Coatings*, **143**, 105598.
6. Hung, H. M., Linh, D. K. & Chinh, N. T. 2019. Improvement of the corrosion protection of polypyrrole coating for CT3 mild steel with 10-camphorsulfonic acid and molybdate as inhibitor dopants. *Progress in Organic Coatings*, **131**, 407–416.
7. Wessling, B. 1994. Passivation of metals by coating with polyaniline: corrosion potential shift and morphological changes. *Advanced Materials*, **6**, 226–228.
8. Tallman, D. E., Spinks, G. & Dominis, A. 2002. Electroactive conducting polymers for corrosion control. *Journal of Solid State Electrochemistry*, **6**, 73–84.
9. Genies, E. M., Bidan, G. J. & Diaz, A. F. 1983. Spectroelectrochemical study of polypyrrole films. *Journal of Electroanalytical Chemistry*, **149**, 101–113.
10. Ohtsuka. T. 2012, Corrosion Protection of Steels by Conducting Polymer Coating,*International Journal of Corrosion*, **2012**, 915090.
11. Babaei-Sati, R., Parsa, J. B. & Vakili-Azghandi, M. 2019. Electrodeposition of polypyrrole/ metal oxide nanocomposites for corrosion protection of mild steel—A comparative study. *Synthetic Metals*, **247**, 183–190.
12. Hosseini, M., Fotouhi, L. & Ehsani, A. 2017. Enhancement of corrosion resistance of polypyrrole using metal oxide nanoparticles: Potentiodynamic and electrochemical impedance spectroscopy study. *Journal of Colloid & Interface Science*, **505**, 213.
13. Van, V. T. H., Hang, T. T. X. & Nam, P. T. 2018. Synthesis of Silica/Polypyrrole Nanocomposites and Application in Corrosion Protection of Carbon Steel. *Journal of Nanoscience & Nanotechnology*, **18**, 4189.
14. Jadhav, N., Kasisomayajula, S. & Gelling, V. J. 2020. Polypyrrole/metal oxides-based composites/nanocomposites for corrosion protection. *Frontiers in Materials*, **7**, 95.

15. Yeganem, M. & Keyvani, A. 2016. The effect of mesoporous silica nanocontainers incorporation on the corrosion behavior of scratched polymer coatings. *Progress in Organic Coatings*, **90**, 296–303.

16. Gong, Y., Shan, Y., Wu, Y., Wang, L., Liu, X., & Ding, F. 2021. Bond Properties of Carbon Fiber Reinforced Polymer and Corrosion-Cracked Reinforced Concrete Interface: Experimental Test and Nonlinear Degenerate Interface Law. *Materials*, **14**, 5333.

17. Sumi, V. S., Arunima, S. R. & Deepa, M. J. 2020. PANI-$Fe_2O_3$ composite for enhancement of active life of alkyd resin coating for corrosion protection of steel. *Materials Chemistry and Physics*, **247**, 122881.

18. Dehghani, A., Zabihi-Gargari, M. & Majd, M. T. 2022. Development of a nanocomposite coating with anti-corrosion ability using graphene oxide nanoparticles modified by Echium ammonium extract. *Progress in Organic Coatings*, **166**, 106778.

19. Kowalski, D., Ueda, M. & Ohtsuka, T. 2010. Self-healing ion-permselective conducting polymer coating. *Journal of Materials Chemistry*, **20**, 7630–7633.

20. Ohtsuka, T., Nishikata, A., Sakairi, M. & Fushimi, K. 2018. *Electrochemistry for Corrosion Fundamentals*. Springer, Singapore, 140–156.

21. Ryu, H., Sheng, N., Ohtsuka, T., Fujita, S. & Kajiyama, H. 2012. Polypyrrole film on 55% Al–Zn-coated steel for corrosion prevention. *Corrosion Science*, **56**, 67–77.

22. Le, H. N. T., Garcia, B., Deslouis, C. & Le Xuan, Q. 2001. Corrosion protection and conducting polymers: polypyrrole films on iron. *Electrochimica Acta*, **46**, 4259–4272.

23. Hien, N. T. L., Garcia, B., Pailleret, A. & Deslouis, C. 2005. Role of doping ions in the corrosion protection of iron by polypyrrole films. *Electrochimica Acta*, **50**, 1747–1755.

24. Schaftinghen, T. V., Deslouis, C., Hubin, A. & Terryn, H. 2006. Influence of the surface pre-treatment prior to the film synthesis, on the corrosion protection of iron with polypyrrole films. *Electrochimica Acta*, **51**, 1695–1703.

25. Kowalski, D., Ueda, M. & Ohtsuka, T. 2007. Corrosion protection of steel by bi-layered polypyrrole doped with molybdophosphate and naphthalene disulfonate anions. *Corrosion Science*, **49**, 1635–1644.

26. Kowalski, D., Ueda, M. & Ohtsuka, T. 2007. The effect of counter anions on corrosion resistance of steel covered by bi-layered polypyrrole film. *Corrosion Science*, **49**, 3442–3452.

27. Kowalski, D., Ueda, M. & Ohtsuka, T. 2008. The effect of ultrasonic irradiation during electropolymerization of polypyrrole on corrosion prevention of the coated steel. *Corrosion Science*, **50**, 286–291.

28. Sasaki, M., Hyono, A., Ohtsuka, T. & Ueda, M.2013, 2013 Winter Meeting of Chemical Society of Japan, Hokkaido Branch, Sapporo.

29. Sasaki, M., Hyono, A., Ueda, M. & Ohtsuka, T. 2012. Corrosion protection of steel by conducting polypyrrole film doped with poly-acids of Mo and W. *ECS Meeting Abstracts*, **20**, 2095.

30. Otero, T. F. & Boyano, I. 2006. Characterization of polypyrrole degradation by the conformational relaxation model. *Electrochimica Acta*, **51**, 6238–6242.
31. Alumaa, A., Hallik, A., Sammelselg, V. & Tamm, J. 2007. On the improvement of stability of polypyrrole films in aqueous solutions. *Synthetic Metals*, **157**, 485–491.
32. Debiemme-Chouvy, C. & Tran, T. T. M. 2008. An insight into the overoxidation of polypyrrole materials. *Electrochemistry Communications*, **10**, 947–950.
33. Marchesi, L. F. Q. P., Simões, F. R., Pocrifka, L. A. & Pereira, E. C. 2011. Investigation of polypyrrole degradation using electrochemical impedance spectroscopy. *Journal of Physical Chemistry B*, **115**, 9570.
34. Cysewska, K., Virtanen, S. & Jasiński, P. 2016. Study of the electrochemical stability of polypyrrole coating on iron in sodium salicylate aqueous solution. *Synthetic Metals*, **221**, 1–7.

# CHAPTER 3

# Inhibition of Copper Corrosion Using Conducting Polypyrrole Coatings

Yanhua Lei and Bochen Jiang

## CONTENTS

DOI: 10.1201/9781003376194-3

## 3.1 CORROSION OF COPPER AND COPPER ALLOYS

Copper alloys are widely used in several industry applications, such as automotive, electronics, fuel gas, telecommunications, and marine.

The good corrosion resistance of these materials may be understood in two ways. Firstly, in an acidic medium, the standard potential of Cu|Cu(I) is more positive than the hydrogen evolution potential. Copper is therefore located in the immunity region. Secondly, in a neutral medium, a uniform and adherent film formed at the metal surface by corrosion products acts as a barrier layer against an aggressive medium. In spite of this self-protective effect, copper and copper alloys may undergo damage in different situations. For instance, in an aerated chloride medium, the corrosion of copper takes place at a noticeable rate. The presence of certain pollutants, such as chloride, sulphides, or ammoniac in the seawater may promote further corrosion of cooling systems.[1,2] According to widespread use of copper in different industries, the issue of corrosion and corrosion protection of copper has attracted a lot of attention and many studies have been conducted to date on this issue and are still ongoing. A schematic illustration of different copper-based industries suffering from corrosion attacks is presented in Figure. 3.1.[1]

In an aqueous solution, Cu will oxidize to either Cu(II) or Cu(I). The latter is only slightly soluble in water, and therefore a film of $Cu_2O$ is the predominant insoluble product during Cu corrosion, while $Cu^{2+}$ is the predominant soluble species.

Two-layer cuprite ($Cu_2O$) films consisting of a compact, epitaxial grown inner layer and a porous outer layer were previously proposed by Ives and Rawson.[3] In addition to cuprite ($Cu_2O$), precipitated Cu(II) species such as $Cu(OH)_2$ and malachite ($Cu_2CO_3(OH)_2$) may occur in an outer layer. Strehblow et al. showed that the Cu passive layer consists of a duplex structure of oxides, with an inner cuprous oxide and an outer cupric hydroxide.[4] Figure 3.2 shows the hypothetical structure of the $Cu/Cu_2O/$ electrolyte interface based on the Ives and Rawson model.[5] One-electron and two-electron reaction mechanisms for Cu dissolution and film formation have been proposed in the literature and are illustrated in

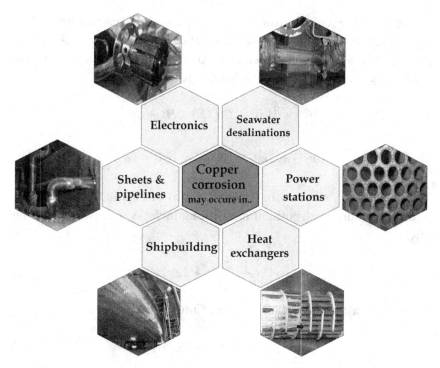

FIGURE 3.1 Schematic illustration of different copper-based industries that suffer from corrosion attacks.

Source: *Arabian Journal of Chemistry* (Open Access).

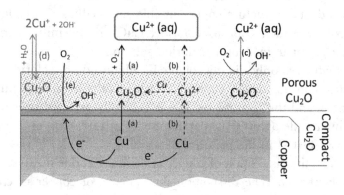

FIGURE 3.2 Mechanisms of film formation at the Cu/cuprite/electrolyte interface.

Source: Corrosion Science (Elsevier).

Figure 3.2. The one-electron reaction mechanism (mechanism (a) in Figure 3.2) involves the stepwise formation of slightly soluble cuprite ($Cu_2O$), followed by the subsequent oxidation of cuprite into a soluble Cu(II) species, as shown in the following reactions:

$$2Cu + H_2O \rightarrow Cu_2O + 2H^+ + 2e^- \tag{3.1}$$

$$Cu_2O + 2H^+ \rightarrow 2Cu^{2+} + H_2O + 2e^- \tag{3.2}$$

The two-electron mechanism (mechanism (b) in Figure 3.1), also known as the dissolution–redeposition mechanism, involves the direct formation of $Cu^{2+}$

$$Cu \rightarrow Cu^{2+} + 2e^- \tag{3.3}$$

followed by a redox reaction between Cu(0) and Cu(II).

$$Cu + Cu^{2+} + H_2O \rightarrow Cu_2O + 2H^+ \tag{3.4}$$

In the presence of $Cl^-$, or other complexing agents, the solubility of Cu(I) is enhanced and becomes the dominant species in solution for example in the form of $CuCl^{-2}$.[2,4,6]

Bengough et al.[7], in 1920, proposed a general stratification of corrosion products on mature copper in chloride medium,[8] as shown in Figure 3.3, in which cuprous chloride, CuCl, is the initial corrosion product of copper in neutral chloride solutions. The formed cuprous chloride, which is only slightly soluble in dilute sodium chloride, reacts to produce cuprous oxide (cuprite), which was the main constituent of thick scales, and the cuprous oxide was generally oxidized over time to cupric hydroxide ($Cu(OH)_2$), atacamite ($Cu_2(OH)_3Cl$), or malachite ($CuCO_3 \cdot Cu(OH)_2$) in the presence of seawater.

After that, the dissolution of copper in neutral chloride media has been extensively studied. The dissolution mechanisms of copper in chloride medium can be represented by the following a two-step mechanism[7,9,10]:

$$Cu + Cl^- \leftrightarrow CuCl_{ads} + e^- \tag{3.5}$$

$$CuCl_{ads} + Cl^- \leftrightarrow CuCl_2^- \tag{3.6}$$

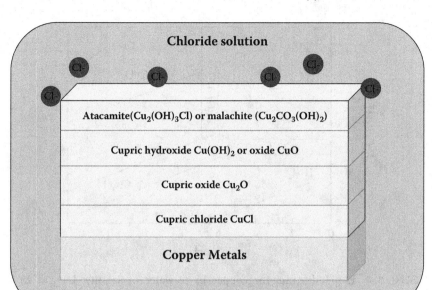

FIGURE 3.3 General stratification scheme of species in a mature copper corrosion product film in seawater, proposed by Bengough et al. in literature.

**Source: created by author for this publication.**

$$CuCl_2^- + Cl^- \leftrightarrow CuCl_3^{2-} + Cl^- \leftrightarrow CuCl_4^{3-} \qquad (3.7)$$

Later, Lee and Noble[11] indicated that cuprous chloride complexes $CuCl_3^{2-}$ and $CuCl_4^{3-}$ will be produced sequentially from $CuCl_2^-$ (reaction (3.7)) as chloride ion concentrations become successively greater than 1.0 mol dm$^{-3}$:

$$2CuCl_2^- + OH^- \leftrightarrow Cu_2O + H_2O + 4Cl^- \qquad (3.8)$$

The cuprous oxide production in the presence of the chloride ion is usually taken as a precipitation reaction according to the reaction of (3.8)[12], rather than a direct electrochemical or chemical formation from the base metal or cuprous chloride.

Bianchi and Longhi investigated the thermodynamic stability of copper in a 3.5% salinity seawater,[13] and produced a number of equilibrium potential-pH (Pourbaix) diagrams for the $Cu/H_2O/Cl^-$ system that included the influence of carbonate and bicarbonate ions. An example has been reproduced in Figure 3.4, where the solid phases taken into consideration are Cu, $Cu_2O$, CuCl, and $Cu_2(OH)_3Cl$. [14]

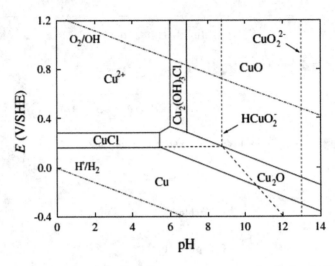

FIGURE 3.4   E–pH diagram for copper in seawater of salinity 3.5% at 25°C (assuming an activity of $10^{-4}$ cuprous and cupric ions in solution.

**Source: Corrosion Science (Elsevier).**

The anodic polarization behavior of pure copper in a chloride medium has received considerable attention in literature. The mechanisms of anodic dissolution of copper in a chloride medium can be represented by the reactions of (3.5) and (3.6), via the formation of $CuCl_{ads}$ and $2CuCl_2^-$.

The anodic polarization of copper at large overpotentials in chloride electrolytes results in E vs. log i curves typified by the schematic shown in Figure 3.5. Three distinct regions that appeared were the active dissolution region (apparent Tafel region), the active-to-passive transition region, and the limiting current region.[15]

**Section III:** a potential above which any increase in current density corresponds to the formation of Cu(II) species.

**Section II:** a potential window of film formation leading to a maximum peak current density and subsequent film or metal dissolution giving a limiting current density.

**Section I:** a potential region of apparent Tafel behavior where mixed charge transfer and mass transport controlling kinetics are usually assumed.

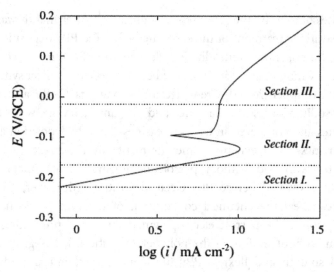

FIGURE 3.5  Anodic polarization curves for copper in chloride solutions.

**Source: Corrosion Science (Elsevier).**

The current first increases and then tends towards a plateau. This indicates a diffusion-limiting rate, attributed to the mass transport of chloride ions (Cl-) to the interface and the diffusion of $CuCl_2^-$ in the solution following the mechanism indicated in Eqs. (3.5) and (3.6).[9] The cathodic reaction for copper in an aerated neutral chloride solution is the oxygen reduction reaction according to the following reaction:

$$O_2 + 2H_2O + 4e^- \rightarrow 4OH^- \tag{3.7}$$

## 3.2  A GENERAL INTRODUCTION OF CONDUCTING POLYMER COATINGS IN COPPER CORROSION PROTECTION

Conducting polymers have been used extensively in the corrosion protection of iron and iron-based alloys.[16–21] Initial work exploring corrosion inhibition of copper alloys using conducing polymer was performed by C.B. Breslin[22] in 2002. A PPy film was formed on a Cu electrode from a near-neutral oxalate solution. The growth of these films was facilitated by the initial oxidation of the copper electrode in the oxalate solution to generate a copper oxalate pseudo-passive layer. After that, highly adherent and homogenous PPy films were electrodeposited at copper from a dihydrogen phosphate solution by the same research group.[23] After that, it was found that passivation treatment of the substrate before the PPy deposition could result in homogenous and adherent deposits of PPy film on copper.[24] K.I. Chane-Ching et al.[25] in 2007, studied the electrochemical synthesis of PPy film on copper from different electrolyte solutions. Oxalic acid, sodium oxalate, sodium/potassium tartrate, and sodium salicylate were used as electrolytes and compared in terms of easiness for PPy deposition and film characteristics. Electrochemical measurements in the presence of these electrolytes with and without pyrrole demonstrate that copper can be efficiently passivated before PPy electrodeposition. Moreover, FT-IR experiments reveal the chemical constitution of the thin passivating layer formed prior to the pyrrole electropolymerization and preventing copper corrosion without inhibiting the polymer formation. J. liesegang et al.[26] obtained smooth and flexible PPy films by electrochemical oxidation of pyrrole on an electropolished Cu electrode in an aqueous dodecylbenzene sulfonic acid (DBSA) medium. It was found that the electropolished copper surface was partially passivated by $Cu_2O$. The interfacial layer between the PPy(DBSA) film and Cu substrate was also confirmed by time-of-flight secondary ion mass spectrometry (TOF-SIMS). Lei et al. synthesized a uniform, compact, and corrosion-resistant PPy coating on copper from a phytic acid solution by constant current oxidation.[27] Protective PPy coatings were then formed on the surface of copper electrode from mixed oxalic acid and DBSA solutions.[28]

To further enhance the protective performance of PPy films, varied inorganic nanoparticles were selected to be dispersed into the electrolytes for performing a PPy electrodeposition film. Then PPy-nano-composited films were obtained directly. Nanoparticle species such as $TiO_2$, copper, silver, carbon nanotube, and graphene were utilized to strength the capacity of the PPy films. For example, CNT incorporated protective PPy-DBS film (named PPy-DBSCNT) were formed at copper by Carmel B. Breslin[29]. The addition of the CNTs was reported to affect the polymer deposition rate as well as the coating's performance in corrosion protection related to inorganic nanoparticles and PPy composited coatings were reviewed in server reports.[30,31]

In addition to using PPy film for copper alloy corrosion, other conductive polymers were also reported to show good corrosion resistance against corrosion. In 2005, Patil et al.[32] reported the synthesis of a poly (o-toluidine) (POT) coating for corrosion protection of copper in a chloride solution, and the results clearly reveal that the POT as a corrosion protective coating on Cu reduces the corrosion rate of Cu almost by a factor of 40. After that in 2006, the authors synthesized the poly(2,5-dimethylaniline) coatings on copper from a salicylate solution.[33] The corrosion rate of copper aqueous 3% NaCl was reduced by a factor of 31 due to the coating of poly(2,5-dimethylaniline). In 2007, poly(o-ethoxyaniline) (POEA) coatings were synthesized on copper by electrochemical polymerization of o-ethoxyaniline (OEA) in an aqueous salicylate solution by using cyclic voltammetry. It was demonstrated that the corrosion potential was about 0.330 V versus *SCE* and more positive in an aqueous 3% NaCl for the POEA coated Cu than that of uncoated Cu and reduced the corrosion rate of Cu almost by a factor of 140.[34]

Poly(o-ethylaniline) coatings were synthesized on copper (Cu) by electrochemical polymerization of o-ethylaniline in an aqueous salicylate solution. The corrosion rate of POEA-coated Cu is found to be 70 times lower than that observed for uncoated Cu. [35] In 2020, A. M. Fathi et al.[36] electrosynthesized protective conducting poly(1,5-diaminonaphthalene) (PDAN) film on copper using a potentiostatic technique. The potentiodynamic polarization and electrochemical impedance spectroscopy measurements proved that the PDAN coating acted as a protective layer on the copper substrate and protected it from corrosion in a 0.6 M NaCl solution.

CPs composted of bilayers were also reported to improve the anticorrosion capacity of the CP films. G. Bereket et al.[37] studied the poly(pyrrole-*co*-N-methyl pyrrole) copolymer and poly(pyrrole)/poly(N-methyl pyrrole) bilayer composites for corrosion protection of copper in a $H_2SO_4$ solution. It was found that copolymer and bilayer coatings were found to have a higher protection effect than single PPy coatings. The same results were obtained by C. B. Breslin when he investigated the anti-corrosion performance of PPy bilayers composed of PPy–Tartrate (PPy-Tra) outside and inner PPy doped dodecylbenzene sulfonate (PPy-DBS) in 2018.[38]

In 2020, Fan et al. electropolymerized the nanostructured PANI, poly(N-ethylaniline) (PNEA) and their molybdate-doped coatings (PANi-Mo and PNEA-Mo) were on passivated copper surface via voltammetric cycles in oxalic acid solution.[39] The potentiodynamic polarization and electrochemical impedance spectroscopy measurements proved that the PANI and PNEA coatings acted as a protective layer on the copper substrate. Doping molybdate is prone to improve the aggregation behavior for PANi and PNEA coatings on copper surface through heterogeneous nucleation with the augment in thickness. Owing to the physical barrier of substituent and in-situ dissolving of molybdate, the corrosion of coated specimens was highly suppressed in $H_2SO_4$ solution. The protective effect of coatings followed the sequence of PNEA-Mo > PANi-Mo > PNEA > PANi with an inverse trend in porosity index.

Although, many efforts have been done to protect copper or copper alloys by conducting polymers, it is still a challenge to synthesize protective and durable conducting coatings on copper. Herein we discuss the electrodeposition process of CPs as well as the potential factors affecting the polymer deposition thorough several examples of PPy electrodeposition on copper.

## 3.3 SYNTHESIS OF PROTECTION CONDUCTING PPY COATINGS FROM PHYTIC ACID SOLUTION

1, 2, 3, 4, 5, 6 Hexakis (di-hydrogen phosphate) myo-inositol (Figure 3.6), known as phytic acid ($IP_6$), was first identified in 1855. Its salt, phytate, is widely present in nature (plants, animals, and soils), mainly as calcium, magnesium, and potassium mixed salts. The salts protect seeds against oxidative damage during storage. Phytic acid and phytate are precipitated by reactions with polyvalent cations. Due to their attractive advantages, including low cost, availability, and nontoxicity, phytic acid and its salts have been widely used in many industrial areas, such as food, medicine,

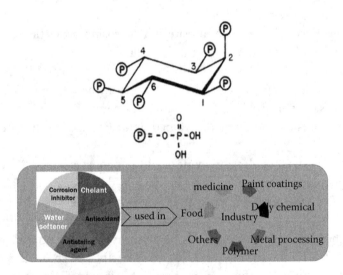

FIGURE 3.6  Structure of phytic acid ($IP_6$ or Phy).

**Source: created by author for this publication.**

and painting. The highly chelating potential of phytate also provides anti-corrosive properties against metal corrosion.[27]

Electropolymerization was performed under constant current control in a three-electrode electrochemical cell with a volume of approximately 50 cm³. The potential of the working electrode was measured with respect to Ag/AgCl/sat. KCl (SSE) and the counter-electrode was platinum foil.

Figure 3.7 shows the potential change as a function of electric charge passed of the copper electrode during constant current polymerization of PPy in a phytic acid solution (pH = 4) containing a Py monomer. The potential transient during the polymerization was characterized by two stages. At the first stage, the potential initially increased from the immersion potential of −0.67 V versus SSE to a plateau around of 0.2~0.3 V. At the second stage, the potential sharply rose to a peak and then slowly decreased to a plateau at which black PPy was observed on the copper surface. The initial potential plateau might relate to the copper dissolution, while the potential peak corresponds to the nucleation of PPy. The charge passed to the peak potential was approximately 0.5 C cm$^{-2}$ in this electrodeposition condition. The three-dimensional growth of PPy occurred in

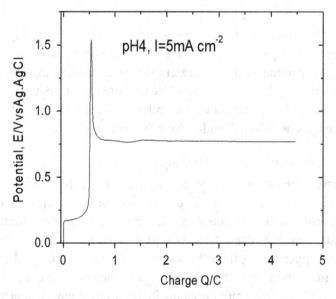

FIGURE 3.7  Potential change with charge during electropolymerizing of PPy on copper in a pH 4 solution of 0.1 M phytic acid and 0.5 M Py monomer.

Source: created by author for this publication.

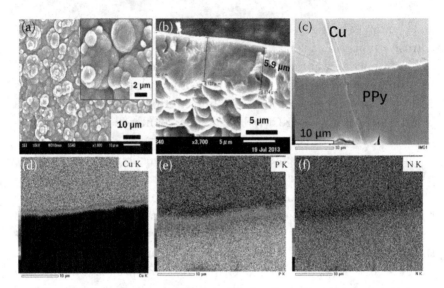

FIGURE 3.8 (a, b) SEM micrographs of PPy-phytate films formed at 5 mA cm$^{-2}$ with 3 C cm$^{-2}$ in a pH 4 solution of 0.1 M phytic acid and 0.5 M Py monomer. (c) Cross-section SEM micrograph of a PPy-phytate film formed on copper with 5 C cm$^{-2}$ charged, (d–f) EDS mapping results of the selected cross-section area.

**Source: created by author for this publication.**

the subsequent potential plateau region. A cauliflower morphology of PPy was observed in Figure 3.8, in which PPy-phytate layers were composed of compactly conglomerated PPy clusters with sizes in the range 5~10 mm. EDS mapping results indicate the successful doping of phytate anions in the PPy matrix. In addition, a transition layer, which is composed of P and Cu elements, was detected under the PPy coatings.

## 3.3.1 EQCM Study of PPy Deposition

The electrochemical quartz crystal microbalance (EQCM) is a powerful method of surface analysis, which can be used for in-situ characterization of the deposition, dissolution and mass changes in the thin film.[40–43]

To study the PPy deposition, EQCM was used to monitor the mass change of copper during the PPy formation. Figure 3.9a shows the EQCM experimental procedure. The mass change of the copper electrode during PPy polymerization was monitored in-situ by electrochemical quartz crystal microbalance (EQCM), in which a frequency sensor, Maxtec KPS550, was coupled with a frequency counter, Advantest 5381. An Au coated AT-cut quartz crystal at a 5 MHz resonance frequency mounted in the frequency

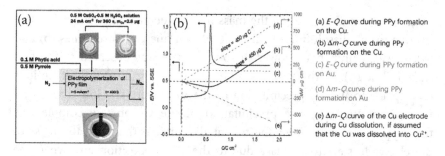

FIGURE 3.9  (a) a schematic of the EQCM experimental procedure, (b) the mass and potential changes of the copper electrode during PPy polymerization.

**Source: created by author for this publication.**

sensor was used as a work electrode. Before the PPy deposition, copper was electrodeposited on the Au-coated quartz crystal in a solution containing 0.5 M $H_2SO_4$ and 0.5 M $CuSO_4$ at 24 mA cm$^{-2}$ for 360 s. The copper deposited was about 2.8 mg cm$^{-2}$, corresponding to a thickness of about 3 μm. The EQCM measurements were performed in a three-electrode cell with a volume of 250 cm$^3$. The mass change ($\Delta m$) was estimated from the frequency change by the Sauerbrey equation. The equation is simplified to the following:

$$\Delta m = -K\Delta f \qquad (3.8)$$

The value of K was determined for a 5 MHz resonance frequency at K = 23.5 ng cm$^{-2}$ Hz$^{-1}$.

Figure 3.9b shows the mass change of the copper electrode during PPy polymerization in the phytic acid solution at pH 4. For comparison, the mass changes of the Au electrode during PPy deposition in the phytic solution was also plotted. The lines (a) and (b) indicate the potential and mass change of the copper electrode during PPy polymerization, respectively; the lines (c) and (d) indicate the potential and mass change of the Au electrode during PPy polymerization, respectively; and line (e) was the mass change of the copper electrode due to the copper dissolution, if assumed that the copper was dissolved into $Cu^{2+}$. As shown in line (c), the mass of the Au electrode increased lineally as the function of charge pass once the current was applied to the Au electrode. When the PPy was deposited on copper, the mass of the copper electrode was changed in accordance with the stages, as shown in line (b). According to the first stage of potential in line (a), the mass initially decreased as shown by line (b) due to the dissolution of substrate. When one compared lines (b) and (d), it was found that a lower dissolution rate was observed at the first stage in line (b). With the dissolution of copper, accumulation of Cu ions takes place to induce precipitation of the salt layer. The low dissolution rate was induced by the precipitation of the Cu-phytate complex on the copper electrode. After the potential exhibited the peak, the mass of the electrode start to increase due to the PPy deposition and when the potential reached the stable plateaus, the mass was linearly increased as shown in line (b) with the same slope of line (d).

### 3.3.2 In-Situ Raman Spectroscopy Study of the PPy Formation

Figure 3.10 compares the typical Raman spectra of the PPy film synthe-sized from the solution of phytic acid and aqueous phytic acid at pH 6[31].

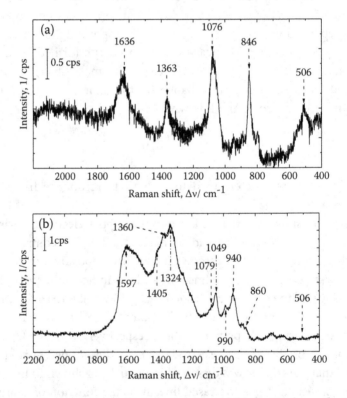

FIGURE 3.10   Raman Spectra of (a) phytic acid, and (b) the PPy layer, the PPy was formed in a phytic acid soltuion at a current dentisy 1.0 mA $cm^{-2}$ for 2,000 s.

**Source: Corrosion Science (Elsevier).**

TABLE 3.1    The Assignments of the Raman Peaks

| Experimental wave numbers/cm$^{-1}$ | Assignments |
| --- | --- |
| 1597 | conjugative C-C=C stretching |
| 1405 | C-N stretching |
| 1324 | ring stretching |
| 1049 | N-H in-plane deformation |
| 990 | ring deformation |
| 940 | C-H out-of-plane stretching |
| 506, 846 | -O-P-OH stretching |

*Source:* Created by authors.

The Raman peaks of the PPy-phytate layer in Figure 3.10(b) were assigned as follows: the peak at 1,597 cm$^{-1}$ to the stretching of conjugative backbone C-C=C-, the peak at 1,405 cm$^{-1}$ to N-C ring stretching, the peak at 1,324 cm$^{-1}$ to C-C ring stretching, and the peak at 1,049 cm$^{-1}$ to -N-H in-plane bending. The peaks at 990 cm$^{-1}$ and 940 cm$^{-1}$ were due to the ring deformation and -C-H out-of-plane deformation. In the Raman spectrum of phytic acid in the solution at pH 6, shown in 10 Figure 3.10(a), five peaks at 506, 846, 1,076, 1,363, and 1,636 cm$^{-1}$ were observed. The peaks at 506 and 846 cm$^{-1}$ were assigned to the -O-P-OH stretching, while the peaks at 1,076 cm$^{-1}$, 1,363 cm$^{-1}$, and 1,636 cm$^{-1}$ were assigned to -P=O deformation, C-H deformation, and H$_2$O bending, respectively.[27] Table 3.1 summarized the assignments of the related peaks.

The in-situ Raman technique is useful in surface analysis. The PPy deposition on copper from a phytic acid solution was further studied by in-situ Raman spectroscopy during electrodeposition. As shown in the following figures, the in-situ Raman spectra were continuously required at different potentials during electro-polymerization (marked on the potential-time curve in Figure 3.11a). And the corresponding results are shown in Figure 3.11b. The peak at 1,147 cm$^{-1}$ is attributed to the Py monomer. As shown in Figure 3.11b, two new peaks appeared at around 933 cm$^{-1}$ and 1,082 cm$^{-1}$ when the potential raised to 0.9 V, duo to the PPy nucleation. According to the previous Raman spectrum of PPy, the two peaks near 933 cm$^{-1}$ and 1,082 cm$^{-1}$ were related to the -C-H out-of-plane deformation and -N-H in-plane deformation, respectively. After that, with the prolonging of electrodeposition, the intensities of the two peaks were obviously enhanced according to the PPy continuous growth.

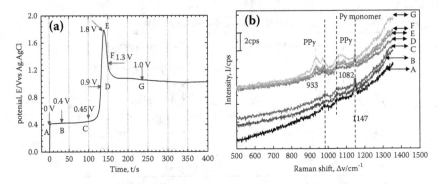

FIGURE 3.11  (a) A potential transition of Cu during the PPy formation (Polymerization were performed at 5 mA cm$^{-2}$ in 0.1 M phytic acid solution with pH 4 and 0.5 M Py monomer), (b) in-situ Raman spectra of the PPy deposition on copper substrate (The Raman spectra were required at different charge that were marked on the Potential-transition curve in (a).).

**Source: created by author for this publication.**

### 3.3.3 Ex-Situ SEM Observation of the PPy Formation

Further, ex-situ SEM was also used to observe the PPy formation on copper. Figure 3.12a shows a potential curve during PPy deposition. The PPy polymerization was interrupted at various predetermined potential values (marked on the curve in Figure 3.12a), and then the surface micromorphology of the sample was observed by SEM. A smooth surface was observed before the addition of current charge for PPy polymerization. A rough surface was observed after the initial charge when the potential increased to 0.2 V. After that, as the potential continuous increased to 0.9 V, small PPy particles were observed on the rough surface. After the potential peak, the potential then slowly decreased to a plateau. After the potential peak, the surface was covered with small globular particles (Figure 3.12e), and with further charge, the globular particles gathered to form larger cauliflower-shaped particles.

Based on the results of in-situ EQCM and in-situ Raman, a schematic representation of the process of Py electropolymerization on copper by using a galvanostatic method from phytic acid was proposed in Figure. 3.12g. During the initial charge, Cu-phytate complex precipitates on the copper surface according to the following equations:

$$Cu \rightarrow Cu^{2+} + 2e^-$$  (3.8)

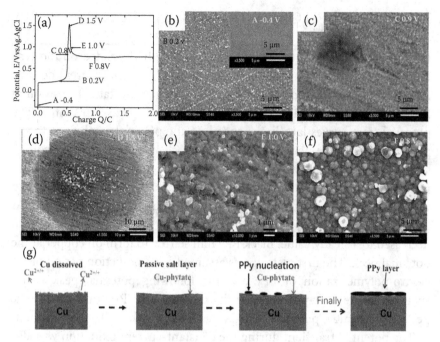

FIGURE 3.12  (a) A potential curve with charge during electropolymerizing of PPy on copper from phytic acid solution (pH =4, I= 5mA cm$^{-2}$). (b–f) The SEM morphology of the sample interrupted at predetermined potential values (corresponding to the potential marked on the curve in Figure 3.12(a)) during PPy polymerization. (g) Schematic representation of the process of PPy electropolymerization on the copper by using galvanostatic method.

**Source: created by author for this publication.**

$$C_6H_{12}O_{24}P_6^{6-} + nCu^{2+} \rightarrow Cu_n(C_6H_{12}O_{24}P_6)^{(6-2n)-} \qquad (3.9)$$

The Cu-phytate complex layer is assumed to function as a passive layer to inhibit the dissolution of copper during the polymerization of PPy. Phytic acid forms a Cu()-phytate complex layer on the copper surface before PPy nucleation begins. Thus, the nucleation of PPy may begin on the surface covered by the Cu(II)-phytate complex layer. The final structure after polymerization is assumed to consist of the thin inner layer of Cu(II)-phytate and the relatively thick outer layer of PPy-phytate.

## 3.4 EFFECT OF PH OF PHYTIC ACID SOLUTION ON THE PPY FORMATION

Interesting, the PPy coating deposition was greatly depended on the synthesized conditions, for example, the pH of the phytic acid, current density. Figure 3.13 shows the potential-time curves for galvanostatic deposition of PPy in the phytate solution at various pH values at CD of 5 mA cm$^{-2}$. The potential-time curves largely depended on the pH of the solution. In the pH from 4 to 6, the potential transient was characterized by the two distinct stages. At the first stage in the transient, the potential revealed a plateau at about 0.2 V, in which the potential of the copper was oxidized to Cu$^{2+}$, although the period and potential depended on the solution pH. At the end of the plateau, the potential gradually rose to a peak and then decreased to a steady value. Black PPy film started to form on copper at the potential peak. The first plateau was assumed as an induction period of the electro-polymerization of PPy and the following potential peak corresponded to the over-potential for the nucleation of PPy on copper. The period of the initial potential plateau was shorter with the higher pH value.

The potential transient during the constant-current oxidation was also influenced by the CD applied. The effect of CD is seen in Figure 3.13b, in which the potential is plotted as a function of charge passed during constant current oxidation at CDs from 1 to 10 mA cm$^{-2}$ in the phytic

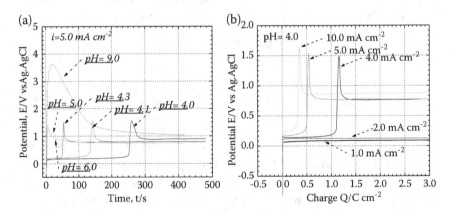

FIGURE 3.13 (a) Potential change with time during electropolymerization of PPy films on copper from the phytic acid solution at different pH values at CD of 5 mA·cm$^{-2}$, (b) potential change with time during electropolymerization of PPy films from a phytic acid solution (pH = 4) at various CDs.

Source: Progress in Organic Coatings (Elsevier).

TABLE 3.2   The Formation of PPy Films in a Mixture Solution of 0.1 M Phytic Acid and 0.5 M Py Monomer at Different Applied Current Density and pH Values

| CD (mA cm$^{-2}$) | pH 1 | pH 3 | pH 4 | pH 4.5 | pH 5 | pH 6 | pH 9 |
|---|---|---|---|---|---|---|---|
| 0.5 | × | × | × | × | × | √ | √ |
| 1 | × | × | × | × | √ | √ | √ |
| 2 | × | × | × | √ | √ | √ | √ |
| 4 | × | × | √ | √ | √ | √ | √ |
| 5 | × | × | √ | √ | √ | √ | √ |
| 10 | × | √ | √ | √ | √ | √ | √ |

Source: Created by authors.
×: The potential does not rise to a high enough potential to polymerize pyrrole.
√: The PPy can be formed in the applied condition.

acid solutions with pH = 4. The potential did not reach the second stage at a CD lower than 2 mA cm$^{-2}$ and no black PPy layer was seen on copper. For the formation of the PPy layer, a CD higher than 4 mA cm$^{-2}$ was required. Table 3.2 summarizes the effects of the solution pH and applied CDs on the formation of the PPy layer. In the solution at pH 1, no polymerization and deposition of PPy occurred in the whole CD. By contrast, the PPy layer was formed in the solutions at pH 6 and 9, even at the small CD of 0.5 mA cm$^{-2}$.

During the PPy deposition from phytic acid solution, a Cu(II)-phytate complex layer was initially formed on copper before the PPy nucleation. The PPy nucleation was started after the initial plateau, in which the Cu-phytate complex surface layer was formed. The larger initial charge pass before the PPy nucleation would result in a thicker Cu-Phytate complex layer. In the acidic solution at pH 1, however, copper underwent heavy dissolution and did not form any surface layer and the potential did not rise to a value high enough to start the polymerization of PPy. In the alkaline solution at pH 9, copper hardly dissolved and did not form the Cu-phytate complex layer. PPy was directly deposited on the copper surface. The instability of the PPy film formed in an alkaline solution may originate from the lack of the intermediate complex layer.

## 3.5 EVALUATION OF CORROSION RATE AND PROTECTIVE EFFICIENCY

### 3.5.1 Open Circuit Potential

Recent studies have shown that a metal substrate coated with a charged conducting polymer will balance the potential of the electrode in the passive region if there is no redox reaction[27]. In addition, the discharge of the conductive polymer film can be seen in the presence of a redox reaction which can be estimated even for the passive zone, which shifts the potential to negative values. The existence of the potential in the passive zone depends on the total charge stored in the polymer and the rate of reaction. The polymer film must therefore be continuously charged for continuous protection. This can be done by the cathodic reduction of the oxygen in the polymer. Thus, open circuit potential (OCP) was general used to estimate the anodic protection property of the charged conducting polymer coating.

The OCP of copper covered with the PPy-phytate layer in a 3.5% NaCl solution was monitored as a function of immersion time. The experimental detail was given in Figure 3.14a, and the result is given in

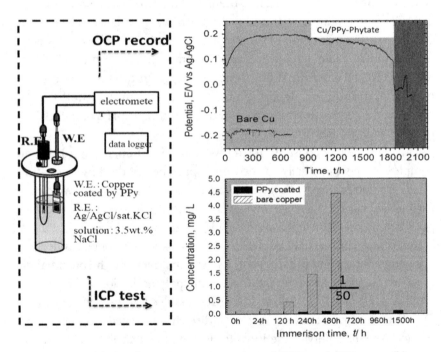

FIGURE 3.14  (a) A schematic representation of OCP and ICP tests, (b) corrosion test open circuit potential (OCP) in a 3.5% NaCl solution of bare copper and copper covered by PPy-phytate film. (c) Amount of Cu ions dissolved from bare copper and copper covered with the PPy-phytate layer during immersion in a 3.5% NaCl solution. (The PPy-phytate layer was formed from phytic acid solution (pH = 4) at 5.0 mA cm$^{-2}$ for 400 s.).

Source: created by author for this publication.

Figure 3.14b, in which the OCP of bare copper is also plotted for comparison. The OCP of copper covered with PPy-phytate initially decreased and then increased to its maximum value at 300 hours. The OCP stayed constant for approximately 600 h and then gradually decreased to −0.05 V versus SSE after 1,800 hours of immersion. For bare copper, the OCP dropped to −0.2 V versus SSE in the initial 3 hours and then remained constant during immersion for 800 hours. On the copper covered with PPy-phytate, no corrosion products were observed in the entire immersion time of 800 hours, although green-colored corrosion products appeared on the bare copper immediately after the immersion began.

## 3.5.2 Amount of Cu Ions Dissolved During Immersion

The concentration of copper ions dissolved in the NaCl solution was measured by inductively coupled plasma-atomic emission spectrometry (ICP-AES). Figure 3.14c shows the concentration of copper ions dissolved from bare copper and copper covered with PPy-phytate during immersion in a 3.5% NaCl solution. For copper covered with PPy-phytate, little Cu ions dissolved in the initial 24 hours of immersion. The concentration of copper ions gradually increased with the immersion time; however, it remained relatively low. The average dissolution rate from the copper covered with the PPy films formed at pH 4 after 480 hours of immersion were about $1.6 \times 10^{-8}$ g cm$^{-2}$ h$^{-1}$, which were much smaller than the average dissolution rate of $8 \times 10^{-7}$ g cm$^{-2}$ h$^{-1}$ for bare copper. That is to say, the corrosion of copper was inhibited by the PPy-phytate film to a factor of 1/50, indicating the excellent corrosion protective performance.

## 3.5.3 Cation Perm-Selectivity of PPy Layer

It was reported that the doped anion played an important role in corrosion protection performance of the conducting polymer coatings. For the redox reaction of a conducting PPy film, the doping behavior is assumed to be changed, depending on the property of the PPy film and doping anions. The mobility of the dopant anions in the PPy is affected by their mass and volume. It was assumed that if the PPy was doped with relative small size anions, as they are easily mobile in the PPy, thus oxidation and reduction of the PPy are accompanied by anions doping from an electrolyte solution and dedoping to an electrolyte solution, respectively. The doping and dedoping process redox reactions could be described in the following equation:

$$PPy^{m+}(m/n)A^{n-} + xe^- \leftrightarrow PPy^{(m-x)+}(m-x/n)A^{n-}$$
$$+ (x/n)A^{n-} \tag{3.10}$$

Reversely, if the PPy was doped with large size anions, the diffusion and migration of the doped ions would be greatly inhibited, as the mobility of the anions with large mass and volume is low in a PPy film, oxidation and reduction of the PPy are accompanied by cations dedoping to an electrolyte solution and doping to the film, respectively, as shown in Eq (3.11).

$$PPy^{m+}(m/n)A^{n-} + xM^+(solution) + xe^- \leftrightarrow PPy^{(m-x)+}(m/n)A^{n-}xM^+$$
$$\tag{3.11}$$

In this case, mobile ions in the PPy should be positively charged particles or cations during oxidation and reduction of the PPy, and thus the transfer particles at the interface of PPy film/aqueous solution should be cations. It influences the protection property of the conducting PPy that the particles transferred at the interface are anions or cations. When anions transfer at the interface and migrate in the film, aggressive chloride ions will easily penetrate to the copper substrate and induce the corrosion of copper. On the contrary, when cations are mobile particles, a chloride attack will be avoided. For understanding the ionic exchange process, two kinds of PPy films were synthesized on copper from a phytic acid and phosphate solution, respectively. EQCM was employed to study the permselective property of the PPy films. The mass change of the PPy film on Au was measured during the PPy redox reaction by EQCM.

For better understanding the ionic exchange process, the mass change in a deposited polymer film on Au during oxidation and reduction processes was in-situ monitored by using the EQCM technique. Figure 3.15a shows the CV curve of PPy-phytate film, while Figure 3.14b gives the mass response from the EQCM electrode coated by PPy-phytate film during the 16th cycle, respectively. A stable CV curve were recorded after the third circle. Usually, when reduction of the PPy film takes place, the anions doped in the PPy film are de-doped out of the PPy film to maintain the neutrality of the polymer, and results in a loss of mass. For the

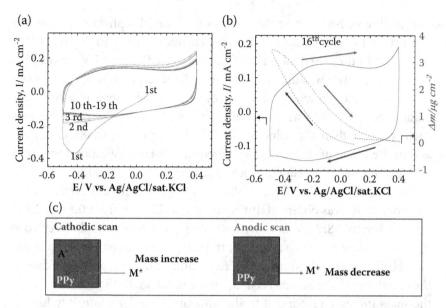

FIGURE 3.15 (a) Cyclic voltammetry plot of the PPy- $phy^{n-}$ in a 3.5 wt.% NaCl aqueous solution, (b) mass change of PPy- $phy^{n-}$ film during cyclic voltammetry. The potential sweep rate was 5 mV s$^{-1}$ in the range of –0.5 to 0.4 V. (c) A schematic of the mass changes of the PPy film during CV.

**Source: created by author for this publication.**

PPy-phytate-coated Au electrode, a mass increase was observed in Figure 3.15b (dashed arrows) during the cathodic scan from an OCP valued at 0.07 to –0.5 V versus SSE. That is to say, instead of the de-doping of anions out of the PPy film, doping of cations (Na$^+$) takes place, during the reduction process of the PPy-phytate film. When one considered the large volume and mass of phytate molecule, the cations such as Na$^+$ were to be de-doped out and doped into the PPy-phytate film to compensate the charged dopants ($phy^{n-}$) immobilized in the polymer layer during the oxidation and reduction of PPy in NaCl solution, respectively. The Na$^+$ ions enter the film, according to the forward reaction of Eq. (3.12) to compensate the negatively charged dopants ($phy^{n-}$ ions) immobilized in the polymer layer:

$$PPy^{m+}(m/n)phy^{n-}_{(doped)} + xNa^+_{(solution)} + xe^- \leftrightarrow PPy^{(m-x)+}(m/n)phy^{n+}xNa^+$$

$$(3.12)$$

The doping of $Na^+$ during the reduction of the PPy-phytate was further confirmed by estimating the Faradic constant value. In Figure 3.15b, according to the potential region from 0.07 (OCP) to −0.5 V versus SSE, the charge density was integrated to be 15.5 mC cm$^{-2}$, resulting in a mass increase by 3.6 μg cm$^{-2}$. Thus, according to the above equation, the Faradic constant (F) was estimated to be 99,155 C mol$^{-1}$. The value was slightly larger than the value of the Faradic constant (96,485 C mol$^{-1}$), and in the range of the errors permitted, it was considered that the mass gain during the PPy reduction was induced by $Na^+$ uptaking.

For the PPy-$H_2PO_4^-$ film, as shown in Figure 3.16, a decrease of the electrode mass was observed during the cathodic scan from 0.1 (OCP) to −0.5 V versus SSE (dashed arrows), and an increase of the electrode mass according to the backward scan process. The mass decrease of the PPy-$H_2PO_4$ electrode during the reduction scan was probably due to the expulsion of the anions ($H_2PO_4^-$ or/ and $Cl^-$) into the solution, while the mass increase responded to the uptaking of anions into PPy film to keep the neutrality of the system as in the following reaction (3.13):

$$PPy^{m+}(m/n)A^{n-} + xe^- \leftrightarrow PPy^{(m-x)+}(m - x/n)A^{n-} + (x/n)A^{n-}$$

$$(3.13)$$

where $A^-$ denoted a $H_2PO_4^-$ or $Cl^-$ anion.

According to the previous results, the de-doping and doping of cations ($Na^+$) took place during the oxidation and reduction of the PPy-phytate film, respectively, due to the large volume and mass of phyhate molecule doped in a PPy matrix. That is to say, the PPy doped with a phytate anion, working as the cationic perm-selective film, restrains the penetration of chloride to the substrate. Different from the behavior of PPy-phytate, however, due to the small volume and mass of the doped $H_2PO_4^-$ in PPy, the doping and de-doping of anions easily took place during the oxidation and reduction of PPy-$H_2PO_4^-$ film, respectively. Thus, when the Cu/PPy-$H_2PO_4^-$ was immersed in the NaCl solution, it was expected that anionic exchanges between the doped $H_2PO_4^-$ and $Cl^-$ in the solution easily takes place. The penetration of chloride ions into a PPy film would accelerate the corrosion of substrate.

### 3.5.4 PPy-Phytate Protection Mechanism

For corrosion protection by a conducting polymer film, two mechanisms have been proposed: anodic protection and physical barrier effect.[27]

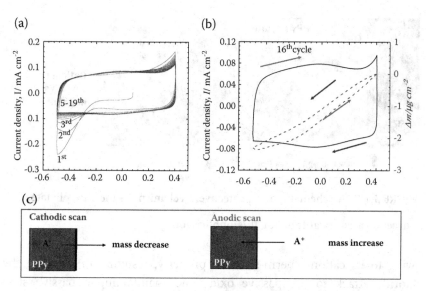

FIGURE 3.16 (a) Cyclic voltammetry plot of the PPy-$H_2PO_4^-$ in a 3.5 wt.% NaCl aqueous solution; (b) mass change of PPy-$H_2PO_4^-$ film during cyclic voltammetry. The potential sweep rate was 5 mV $s^{-1}$ in the range of $-0.5$ to 0.4 V; (c) a schematic of the mass changes of the PPy film during CV.

**Source: created by author for this publication.**

On the former mechanism, the oxidative action of the polymer film induces passivation of the substrate metal and the passive state of the metal is kept under a state of a low dissolution rate. On the latter mechanism, the polymer film works as a barrier against the penetration of oxidants and aggressive anions, protecting the substrate metals.

Based on the results in this investigation, the dissolution of copper in 3.5% NaCl was much restrained by the PPy-phytate film. Figure 3.17 were proposed to illustrate the protection mechanism of the PPy-phytate film. First, due to the oxidative property of the PPy-phytate film, the redox potential of the PPy films is significantly more positive than the corrosion potential of bare copper. The copper covered by PPy-phytate revealed high potentials at about 0.1–0.2 V versus SSE, in which copper should be covered by the passive $Cu_2O$/CuO film. Meanwhile, the oxidative property of the PPy further allows the oxide to be stably maintained. Second, the polymer film works as a barrier against the penetration of oxidants and aggressive anions. Further, a protective phytate-Cu complex layer was formed between the PPy film and substrate. Thirdly, it has been proved that the PPy-phytate film can prevent the penetration of chloride ions

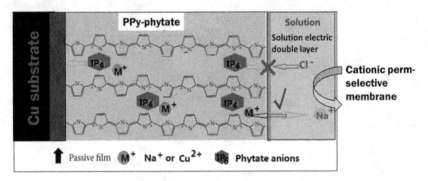

FIGURE 3.17    A schematic of a protective mechanism of the PPy-phytate.

**Source: created by author for this publication.**

owing to its cationic perm-selective property, resulting in avoiding the chloride attack to the passive oxide and maintaining a passive state without dissolution of copper.

## 3.6 EFFECTS OF INHIBITORS ON THE PPY DEPOSITION AND ITS PROTECTION PROPERTIES

With acidic mediums for the PPy deposition, the oxidation of copper occurs before the pyrrole (Py) monomer is oxidized. Due to its low corrosion potential in aqueous media, copper dissolves at the potential that is necessary to oxidize the monomer. In order to deposit a film of conducting polymers, it is essential to work in a medium with an electrolyte that can protect the copper surface from dissolution without impeding the electropolymerization process.[44] When the polymeric PPy film grows with the formation of copper oxide, it was reported that the deposited PPy became less homogeneous and less adherent.[24] For the formation of homogenous and adherent PPy films, pretreatment of the copper substrate, such as by surface passivation[24,44] and formation of a self-assembled organic layer[45] prior to the PPy electropolymerization, has been reported.

Benzotriazole (BTAH) is known as an effective corrosion inhibitor for copper and its alloys in aqueous solutions, and is widely used in industry.[46–48] It is known that a protective barrier layer, consisting of a complex between copper and BTAH, is formed when Cu is immersed in a solution containing BTAH. Researchers have proposed various absorption modes of BTAH on copper.[46,49] For example, a Cu-BTA complex chemisorbed mode on a Cu surface was proposed by Cotton et al.[50] In solutions containing BTA, a protective Cu (I) BTA layer is rapidly formed on copper and copper alloy surfaces[51]. The protective layer thus formed is expected to hinder dissolution of the copper and enable the Py monomer to be easily electropolymerized.

### 3.6.1 BTA Adsorption on Copper (BTA-ad-Cu) in an Oxalic Acid Solution During PPy Electrodeposition

The PPy was electrodeposited from an oxalic acid solution. For the PPy formation, an oxalic acid (Ox) aqueous solution at 0.1 mol dm$^{-3}$ (M) was used, and BTA was added at 0.01 M in the solution. When copper is immersed in a solution containing BTA, a protective adsorption layer is assumed to form and inhibit the copper corrosion. The formation of the BTA layer on copper during the initial 600 s of immersion was examined by EQCM. Figure 3.18 shows the mass change of the copper electrode during the initial 600 s in the solution of oxalic acid and Py monomer containing BTA or not containing BTA[52]. In Figure 3.18, the mass of the copper electrode slowly decreased without BTA. In the presence of BTA, however, the mass gradually increased to 1.5 µg cm$^{-2}$ after 600 s of immersion. The mass increase of the electrode is correspondent to the adsorption of BTA and Cu-BTA complex layer formation.

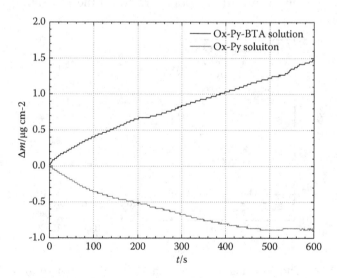

FIGURE 3.18 Mass change of Cu during the initial 600 s of immersion in Ox-Py and Ox-Py-BTA solutions.

**Source: Progress in Organic Coatings (Elsevier).**

### 3.6.2 Effect of the BTA Adsorption on PPy Formation

Due to the adsorption of BTA on copper surface, thus the dissolution of the copper during PPy electrodeposition would be restrained. Then EQCM was also employed to monitor the mass changes of working-electrode during the process of PPy deposition.

The potential change and the mass change ($\Delta m$) of the copper electrode during the PPy polymerization are shown in Figure 3.19(a) in the oxalic acid (Ox)-Py solution with/without BTA addtion[52]. Lines (a) and (b) indicate the potential and mass change, respectively, in the Ox-Py-BTA solution and lines (c) and (d) indicate the potential and mass change, respectively, in the Ox-Py solution. Two stages of the potential transient curve were recorded during the PPy polymerization from Ox-Py solution. At the first stage, the potential revealed a plateau at about 0.2 V. At the second stage, the potential sharply rose to a peak and then slowly decreased to a steady value. The mass was changed in accordance with the stages. The mass initially decreased, and then greatly increased as shown by line (d) in Figure 3.19(a). After the potential exhibited the peak, the rate of the mass increase degraded and at the time period of the second potential plateau at 0.6 V, the mass linearly increased. The potential peak corresponds to the over-potential for nucleation of PPy and the following potential plateau to three-dimensional growth of the PPy film.

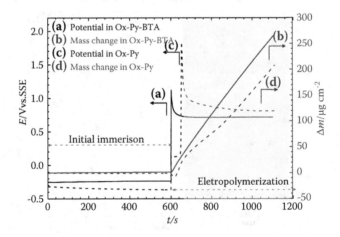

FIGURE 3.19 The potential (curve a) and mass change (curve b) during PPy polymerization in an Ox-Py-BTA solution and the potential (curve c) and mass change (curve d) in an Ox-Py solution.

**Source: Progress in Organic Coatings (Elsevier).**

In the electrolyte with BTA addition, the potential initially increased without the initial plateau from the immersion potential of −0.25 V to a peak and then rapidly decreased to the second potential plateau. The mass linearly increased with increase of time without delay. The linear increase of the mass may indicate that the PPy film uniformly grows with the charge applied. The Cu-BTA complex layer on a copper surface formed during the initial immersion for 600 s plays a large role on the uniform growth of the PPy film on copper. Smooth, homogenous, and adhesive PPy films were obtained from the oxalic acid solution with BTA addition, as shown in Figure 3.20(a,b). Three-dimensional views of the surface of the PPy films, were observed with CLSM, as shown in Figure 3.20(c,d). The mean roughness of the level on the PPy film surface (Ra) for the PPy film formed in the Ox-Py-BTA solution Ra = 0.4 μm and for the film in the Ox-Py solution Ra = 0.8 μm. The addition of BTA results in the formation of a more smooth PPy film.

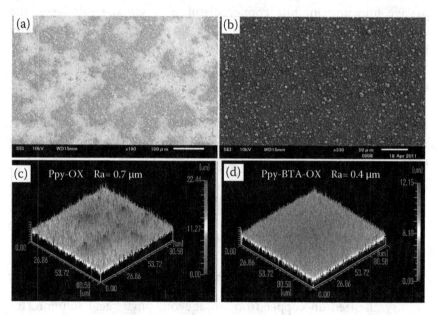

FIGURE 3.20  SEM images of the PPy films formed from (a) Ox-Py solution, (b) Ox-BTA-Py solution; 3-dimesional view of the PPy film formed on copper, (c) Ox-Py solution, (d) Ox-BTA-Py solution (The PPy films were prepared at 2.0 mA cm−2 for 500 s.).

**Source: Progress in Organic Coatings (Elsevier).**

In addition, both the peak and plateau potentials in the Ox-Py-BTA solution were lower than those in the Ox-Py solution in Figure 3.19, indicating the decrease of over-potential nucleation due to the formation of a Cu-BTA complex layer.

Formation of PPy on metal substrate from oxalate solution has been investigated. In the oxalate solution, the PPy layer was reported to be stably formed, probably because initial formation of metal-oxalate salt layer passivated the metal surface and facilitated the subsequent polymerization deposition of PPy. The EQCM measurement provided the useful result on the formation process of PPy in the Ox-Py solution. As shown in curve (d) in Figure 3.19, the mass initially decreased and then sharply increased in the initial potential plateau. The initial decrease of the mass corresponds to dissolution of copper, and the subsequent increase may be related to deposition of copper oxalate salt. With the dissolution of copper, accumulation of Cu ions takes place to induce precipitation of the salt layer. The copper oxalate salt layer is assumed to work as a passive film and inhibits copper dissolution. On the salt layer, the electropolymerization is started and then the mass was linearly increased with the growth of the PPy film.

A protective BTA layer or Cu ion-BTA complex layer was formed on copper during the initial immersion and inhibit dissolution of copper. The polymerization of PPy starts on the protective layer immediately after the anodic current is imposed. When the PPy film was prepared on copper in the oxalic acid solution containing BTA, the following improvements were obtained:

1. Surface of the PPy film was more homogeneous.
2. Adhesion between the copper and the PPy film was stronger.
3. Corrosion protection in the NaCl solution was enhanced.

## 3.7 SUMMARY

1. A polypyrrole (PPy) doped with phytate anions was electro-synthesized on copper from an aqueous phytic acid solution. In-situ Raman and EQCM techniques could be used to study the PPy deposition process. A thin layer composed of a Cu-phytate complex was initially formed, followed by the polymerization of PPy doped with phytate anions. The polymerization process of PPy on copper changed with the pH values of a phytic acid solution and current density (CD) applied.

2. The open circuit potential (OCP) of copper in a 3.5 wt% NaCl solution was shifted to a high potential state by PPy-phytate films. The high-potential state was maintained for longer than 75 days by the PPy film prepared in the phytate solution at pH 4.

3. Two different PPy coatings were coated on copper from a phytic acid and sodium di-hydrogen phosphate solution, respectively. Investigations were performed on the PPy coated copper in 3.5% NaCl by EQCM. The PPy film doped with phytate anions as a cationic selective coating effectively protected the substrate from attack; however, different from the PPy-phytate coating, the PPy doped with di-hydrogen phosphate anions, mainly works as a physical barrier to restrain the diffusion of chloride.

4. Inhibitor of benzotriazole (BTA) was used to pre-treat copper substrate for polypyrrole (PPy) deposition. Homogeneous and adherent polypyrrole (PPy) film was synthesized on the BTA-pretreated copper electrode. The BTA pretreatment facilitated the PPy nucleation and improved the adhesion of the obtained PPy film.

## REFERENCES

1. Fateh, A., Aliofkhazraei, M. & Rezvanian, A. R. 2020. Review of corrosive environments for copper and its corrosion inhibitors. *Arabian Journal of Chemistry*, **13**, 481–544.

2. Rahmouni, K., Keddam, M., Srhiri, A. & Takenouti, H. 2005. Corrosion of copper in 3% NaCl solution polluted by sulphide ions. *Corrosion Science*, **47**, 3249–3266.

3. Ives, D. J. G. & Rawson, A. E. 1962. Copper corrosion I. thermodynamic aspects. *Journal of The Electrochemical Society*, **109**, 447–451.

4. Titze, H. H. & Titzb, B. 1980. The investigation of the passive behaviour of copper in weakly acid and alkaline solutions and the examination of the passive film by esca and ISS. *Electrochimica Acta*, **25**, 839–850.

5. Zhou, P., Ogle, K. & Wandelt, K. 2018. The corrosion of copper and copper alloys. *Encyclopedia of Interfacial Chemistry: Surface and ELectrochemistry*, **6**, 478–489.

6. Adeloju, S. B. & Hughes, H. C. 1986. The corrosion of copper pipes in high chloride-low carbonate mains water. *Corrosion Science*, **26**, 851–870.

7. Khiati, Z., Othman, A. A., Sanchez-Moreno, M., Bernard, M. C., Joiret, S., Sutter, E. M. M., & Vivier, V. 2011. Corrosion inhibition of copper in neutral chloride media by a novel derivative of 1,2,4-triazole. *Corrosion Science*, **53**, 3092–3099.

8. Bengough, G. D., Jones, R. M. & Pirret, R. 1920. Diagnosis of brass condenser tube. corrosion. *Journal of the institute of metals*, 65–158.

9. Khaled, K. F. 2011. Studies of the corrosion inhibition of copper in sodium chloride solutions using chemical and electrochemical measurements. *Materials Chemistry and Physics*, **125**, 427–433.

10. Liu, T., Chen, S., Cheng, S., Tian, J., Chang, X. T., & Yin, Y. S. 2007. Corrosion behavior of super-hydrophobic surface on copper in seawater. *Electrochimica acta*, **52**, 8003–8007.

11. Lee, H. P. & Nobe, K. 1987. ChemInform abstract: kinetics and mechanisms of Cu electrodissolution in chloride media. *ChemInform*, **133**, 2035.

12. Bianchi,G., Fiori, G., Longhi, P. & Mazza, F. 1978. "Horse Shoe" Corrosion of copper alloys in flowing sea water: mechanism, and possibility of cathodic protection of condenser tubes in power stations. *Corrosion*, **34**, 396–406.

13. Faita, G., Fiori, G. & Salvadore, D. 1975. Copper behaviour in acid and alkaline brines—I kinetics of anodic dissolution in 0.5M NaCl and free-corrosion rates in the presence of oxygen. *Corrosion science*, **15**, 383–392.

14. Bianchi, G. & Longhi, P. 1973. Copper in sea-water, potential-pH diagrams. *Corrosion Science*, **13**, 853–864.

15. Kear, G., Barker, B. D. & Walsh, F. C. 2004. Electrochemical corrosion of unalloyed copper in chloride media––a critical review. *Corrosion Science*, **46**, 109–135.

16. Sanad, S. H., Fayyad, E. M. & Ismail, A. A. 2018. Characterization and corrosion protection ability of conducting polymer coatings on mild steel in acid media. *Silicon*, **11**, 1221–1234.

17. Inzelt, G. 2017. Recent advances in the field of conducting polymers. *Journal of Solid State Electrochemistry*, **21**, 1965–1975.

18. Dura, A. & Breslin, C. B. 2019. Electrocoagulation using aluminium anodes activated with Mg, In and Zn alloying elements. *J Hazard Mater*, **366**, 39–45.

19. Deshpande , P P & Sazou, D . 2016.Corrosion Protectioncorrosion protection of Metalsmetals by Intrinsically Conducting Polymersintrinsically conducting polymers.,1st Edition,CRC Press.ISSUE: 9780429090929

20. Deshpande, P. P., Bhopale, A. A., Mooss, V. A. & Athawale, A. A. 2016. Conducting polyaniline/nano-zinc phosphate composite as a pigment for corrosion protection of low-carbon steel. *Chemical Papers*, **71**, 189–197.

21. Deshpande, P., Vathare, S., Vagge, S., Tomšík, E. & Stejskal, J. 2013. Conducting polyaniline/multi-wall carbon nanotubes composite paints on low carbon steel for corrosion protection: electrochemical investigations. *Chemical Papers*, **67**, 1072–1078.

22. Fenelon, A. M. & Breslin, C. B. 2002. The electrochemical synthesis of polypyrrole at a copper electrode: corrosion protection properties. *Electrochimica Acta*, **47**, 4467–4476.

23. Redondo, M. I. & Breslin, C. B. 2007. Polypyrrole electrodeposited on copper from an aqueous phosphate solution: Corrosion protection properties. *Corrosion Science*, **49**, 1765–1776.

24. Herrasti, P., Del Rio, A. I. & Recio, J. 2007. Electrodeposition of homogeneous and adherent polypyrrole on copper for corrosion protection. *Electrochimica Acta*, **52**, 6496–6501.

25. Dos Santos, L. M., Lacroix, J. C., Chane-Ching, K. I., Adenier, A., Abrantes, L. M., & Lacaze, P. C. 2006. Electrochemical synthesis of polypyrrole films on copper electrodes in acidic and neutral aqueous media. *Journal of Electroanalytical Chemistry*, **587**, 67–78.

26. Prissanaroon, W., Brack, N., Pigram, P. J. & Liesegang, J. 2004. Electropolymerisation of pyrrole on copper in aqueous media. *Synthetic Metals*, **142**, 25–34.

27. Lei, Y. H., Sheng, N., Hyono, A., Ueda, M. & Ohtsuka, T. 2013. Electrochemical synthesis of polypyrrole films on copper from phytic solution for corrosion protection. *Corrosion Science*, **76**, 302–309.

28. Menkuer, M. & Ozkazanc, H. 2019. Electrodeposition of polypyrrole on copper surfaces in OXA-DBSA mix electrolyte and their corrosion behaviour. *Progress in Organic Coatings*, **130**, 149–157.

29. Carragher, U., Branagan, D. & Breslin, C. B. 2019. The influence of carbon nanotubes on the protective properties of polypyrrole formed at copper. *Materials*, **12**, 2587.

30. Zadeh, M. K., Yeganeh, M., Shoushtari, M. T. & Esmaeilkhanian, A. 2021. Corrosion performance of polypyrrole-coated metals: A review of perspectives and recent advances. *Synthetic Metals*, **274**, 116723.

31. Ehsani, A., Heidari, A. A. & Sajedi, M. 2020. Graphene and Graphene/polymer composites as the most efficient protective coatings for steel, aluminum and copper in corrosive media: a review of recent studies. *The Chemical Record*, **20**, 467–493.

32. Shinde, V., Sainkar, S. R. & Patil, P. P. 2005. Corrosion protective poly(o-toluidine) coatings on copper. *Corrosion Science*, **47**, 1352–1369.

33. Shinde, V., Gaikwad, A. B. & Patil, P. P. 2006. Synthesis and characterization of corrosion protective poly(2,5-dimethylaniline) coatings on copper. *Applied Surface Science*, **253**, 1037–1045.

34. Chaudhari, S. & Patil, P. P. 2007. Corrosion protective poly(o-ethoxyaniline) coatings on copper. *Electrochimica Acta*, **53**, 927–933.

35. Shinde, V. & Patil, P. P. 2010. Evaluation of corrosion protection performance of poly(o-ethyl aniline) coated copper by electrochemical impedance spectroscopy. *Materials Science and Engineering: B*, **168**, 142–150.

36. Fathi, A. M. & Mandour, H. S. 2019. Electrosynthesized conducting poly(1,5-diaminonaphthalene) as a corrosion inhibitor for copper. *Polymer Bulletin*, **77**, 3305–3324.

37. Cakmakci, I., Duran, B. & Bereket, G. 2013. Influence of electrochemically prepared poly(pyrrole-co-N-methyl pyrrole) and poly(pyrrole)/poly(N-methyl pyrrole) composites on corrosion behavior of copper in acidic medium. *Progress in Organic Coatings*, **76**, 70–77.

38. Carragher. U. & Breslin. C. B. 2018. Polypyrrole doped with dodecylbenzene sulfonate as a protective coating for copper. *Electrochimica Acta*, **291**, 362–372.

39. Ma, Y., Fan, B. M., Liu, H., Fan, G. F. et al. 2020. Enhanced corrosion inhibition of aniline derivatives electropolymerized coatings on copper: Preparation, characterization and mechanism modeling. *Applied Surface Science*, **514**, 146086.

40. Wang, Y., Zhang, Y., Zhou, B., Li, C. & Gao, F. 2019. In-situ observation of the growth behavior of ZnAl layered double hydroxide film using EQCM. *Materials & Design*, **180**, 107952.

41. Ma, Z., Yang, Y., Brown, B., Nesic, S. & Singer, M. 2018. Investigation of precipitation kinetics of $FeCO_3$ by EQCM. *Corrosion Science*, **141**, 195–202.

42. Lei, Y. H., Sheng, N., Hyono, A., Ueda, M. & Ohtsuka, T. 2014. Influence of pH on the synthesis and properties of polypyrrole on copper from phytic acid solution for corrosion protection. *Progress in Organic Coatings*, **77**, 774–784.

43. Sharifirad, M., Omrani, A., Rostami, A. A. & Khoshroo, M. 2010. Electrodeposition and characterization of polypyrrole films on copper. *Journal of Electroanalytical Chemistry*, **645**, 149–158.

44. Liu, Y. C. & Chung, K. C. 2003. Characteristics of conductivity-improved polypyrrole films via different procedures. *Synthetic Metals*, **139**, 277–281.

45. Shustak, G., Domb, A. J., & Mandler, D. 2006. n-Alkanoic Acid Monolayers on 316L Stainless Steel Promote the Adhesion of Electropolymerized Polypyrrole Films. *Langmuir*, **22**, 5237–5240.

46. Lin, J. Y. & West, A. C. 2010. Adsorption–desorption study of benzotriazole in a phosphate-based electrolyte for Cu electrochemical mechanical planarization. *Electrochimica Acta*, **55**, 2325–2331.

47. Finšgar, M. & Milošev, I. 2010. Inhibition of copper corrosion by 1,2,3-benzotriazole: A review. *Corrosion Science*, **52**, 2737–2749.

48. Izquierdo, J., Santana, J. J., González, S. & Souto, R. M. 2010. Uses of scanning electrochemical microscopy for the characterization of thin inhibitor films on reactive metals: The protection of copper surfaces by benzotriazole. *Electrochimica Acta*, **55**, 8791–8800.

49. Fateh, A., Aliofkhazraei, M. & Rezvanian, A. R. 2020. Review of corrosive environments for copper and its corrosion inhibitors. Arabian Journal of Chemistry, **13**, 481–544.

50. Cotton, J. B. in *Proceedings of the 2nd International Congress on Metallic Corrosion*. 590–595.

51. Spah, M., Spah, D. C., Singh, K. C., Gupta, V., Song, H. J. & Park, J. W. 2008. Effect of Solvents on the Corrosion Inhibition of Copper by 1,2,3-Benzotriazole. *Journal of Solution Chemistry*, **37**, 1197–1206.

52. Lei, Y. H., Sheng, N., Hyono, A., Ueda, M. & Ohtsuka, T. 2014. Effect of benzotriazole (BTA) addition on Polypyrrole film formationon copper and its corrosion protection. *Progress in Organic Coatings*, **77**, 339–346.

# CHAPTER 4

# Corrosion Protection of Magnesium (Mg) Alloys by Conducting Polymers

Nan Sheng and Jingxiang Xu

## CONTENTS

## 4.1 INTRODUCTION

Magnesium and its alloys have been utilized in innovative applications in many fields, such as aerospace components, automobiles, computers, etc., due to their desirable properties of low density,[1–4] high specific strength,

DOI: 10.1201/9781003376194-4

and good processing ability. However, disturbed by the poor corrosion resistance and high chemical reactivity of the Mg alloys, practical industrial applications of the Mg alloys are still restricted.[4–7] Thus, the enhancement of comprehensive properties by alloying elements is of great necessity for the Mg and its alloys. Therefore, new surface finishing for the enhancement of corrosion resistance is expected to be developed.

Nowadays, organic coatings have been of interest in the feasible use of corrosion protection of metal, since the conducting polymer coating is one of the green methods for the corrosion protection of metals and has recently attracted considerable attention for the application in corrosion protection. Many papers have been published for corrosion inhibition of metals by conducting polymers (CPs) with coatings such as polypyrrole (PPy), polyaniline (PANI) and polythiophene (PTh).[8–12] It has been assumed that the CPs with oxidative properties allowed the metals to be passivated and facilitated formation of passive oxides on the metals.

For the development of an effective protection layer by using a CP layer, an over-layer of PPy doped with large-size organic anions covered on an inner layer could further improve the corrosion resistance.[13,14] Electropolymerization under ultrasonic irradiation could also improve the anticorrosive property of a PPy layer.[15] The imposition of an ultrasonic wave in a liquid medium allows gas bubbles to generate, which grow and collapse, greatly changing the pressure and temperature inside the bubbles. Various physico-chemical phenomena were induced by the imposition of an ultrasonic wave. The use of an ultrasonic wave is thus expected to induce changes of electro-synthesis conditions for the PPy preparation and to modify the properties of the PPy layer.

In our study, in order to improve the protection property of a PPy layer against corrosion, we finally prepared a bilayered PPy film on the zinc-coated AZ91D alloy.[16,17] This bilayered PPy film contained inner PPy-Tartrate-Molybdate layer and outer PPy-dodecylsulfate layer.[18] The effect of ultrasonic irradiation on electropolymerization of the inner PPy layer was also investigated. Corrosion inhibition properties of the bilayered PPy layer on the zinc-coated AZ91D alloy was tested by immersion in a 3.5 wt.% sodium chloride aqueous solution. The protection time of the bilayered PPy film formed with ultrasonic irradiation can be maintained for 221 hours.

## 4.2 CORROSION BEHAVIOR OF MG ALLOYS

Magnesium is the lightest engineering metal in the world (the density of Mg is 1.7 $g/cm^3$), which is significantly lower than aluminum

(Al, 2.7 g/cm$^3$), titanium (Ti, 4.5 g/cm$^3$), and iron (Fe, 7.9 g/cm$^3$), respectively. In addition, Mg is the eighth most abundant element in earth's crust, which indicated we have sufficient resources to use the Mg alloys in various engineering field. Mg alloys have been shown as one of the highest strength-to-weight ratio of structural alloys; however, other properties like corrosion, ductility, and creep limit its application.

Joseph Black first proposed Mg is an element in 1755.[19] H. Davy et al. isolated the metallic Mg in 1808; after that, A. Bussey prepared metallic Mg in coherent in 1831. The study of the corrosion property of Mg was documented in 1831. A. Bussey did the first corrosion experiment in 1931. A. Bussey carried out the first corrosion experiment on Mg and reported "Mg becomes covered with the hydroxide after exposure to moist air, but remains un-attacked in dry air".[20] W. Beetz did a corrosion study of Mg in an aqueous solution in 1866,[21] who pointed out that Mg exhibits hydrogen evolution during anodic polarization.

In recent years, researchers persistently studied corrosion of Mg alloys to expand their application. M. Esmaily et al. reviewed the key developments in regards to corrosion of Mg alloys and summarized the timeline of scientific and technological developments of Mg corrosion research, as shown in Figure 4.1.[19] M. Ali et al. reviewed the mechanical and corrosion properties of the research on Mg-based biocomposites.[22] This indicated that the Mg-based alloys could be extensively used in medical implants, dentistry, statures, biosensors, bioelectrodes, skin substitutes, and drug delivery systems.[22–26] D. Xu reviewed the influence of the long-period stacking ordered phase on the properties of corrosion resistance and mechanical strength.[27] They also described the unsolved issues of the long-period stacking ordered phase containing Mg alloys and pointed out how to improve the service properties.

## 4.3 CORROSION PROTECTION OF CONDUCTING POLYMERS IN OUR WORK

Organic polymers are generally insulators, and because of the cheap raw materials, lightweight, low temperature processability, and corrosion-resistant properties, the organic polymers have been increasingly used as structural materials similar to ceramics, wood, and metals. As an insulator, the conventional polymers can be modified to blend with pigments to improve electrical, mechanical, and corrosion resistance. By using this blend modified method, the conductivity is not ascribed to the structure

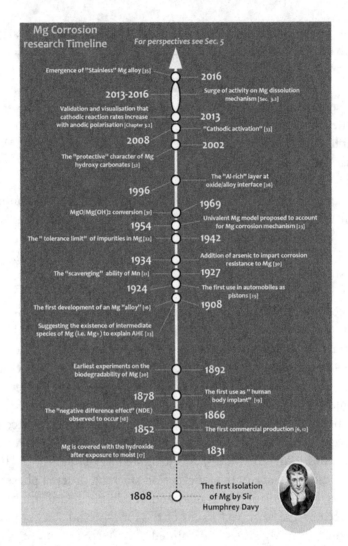

FIGURE 4.1 Timeline of scientific and technological developments of Mg corrosion research, indicating "landmarks" since the isolation of the Mg metal in 1808.

**Source: Progress in Materials Science (Open Access).**

of polymer, in which the conventional polymers are distinguishable from conducting polymers.

Polymers to be classified in the intrinsically conducting polymers must possess a conjugated backbone; the conjugated backbone provides a great degree of delocalization of π-electrons. The electronic conduction was suggested by the movement of the π-electrons from molecule to molecule in the complicated CPs structure, as the CPs have the chemical structure that can produce, sustain, and assist the motion of charge carriers (electrons and holes) necessary for electrical conduction.[28,29] The overlapping of molecular orbitals gives reasonable carrier mobility along the polymer chain.

The experimental research presented in our work is concerned with the CPs as coatings for corrosion prevention of magnesium alloy of AZ91D. For the preparation of PPy coatings, various organic electrolytes can be applied to the electrodeposition of the PPy layer on the oxidizable metals, such as tartrate salt,[30–32], salicylate salt,[33,34] and malic acid.[35,36]

In our study, we successfully prepared a PPy film on a zinc-coated magnesium alloy of AZ91D by electrochemical oxidation under constant current control in a sodium tartrate aqueous solution.[16–18] To improve the corrosion protection property of the PPy film, several methods were investigated. Firstly, a two-step constant current control was used to electropolymerize more compact PPy film[18]: (1) the initial larger current density electropolymerization process and (2) a lower current density process electropolymerization process. Secondly, a new anionic doping salt was introduced into the formation process, with the newly introduced anionic doping salt, a bilayered PPy film was formed.[18] The bilayered PPy film included the inner layer (PPy-tartrate-$MoO_4$ layer) formed under the two-step constant current control and the outer layer (PPy-DS layer) formed under constant current control. Thirdly, the effect of ultrasonic irradiation during Py electropolymerization was investigated.

The formation process of the PPy layer on the zinc-coated AZ91D alloy in a sodium tartrate solution was explained and demonstrated by surface morphology and Raman spectra. The corrosion prevention property was studied by an immersion test in a sodium chloride aqueous solution.

## 4.4 THE PPY FILMS FOR CORROSION PROTECTION OF A ZINC-COATED AZ91D ALLOY

### 4.4.1 The Preparation of the Conducting Polypyrrole Layer

Because the redox potential of Mg and the oxidation potential of the Py monomer exhibited a large difference, direct electro-synthesis of the polypyrrole (PPy) in an aqueous medium on the magnesium alloy of

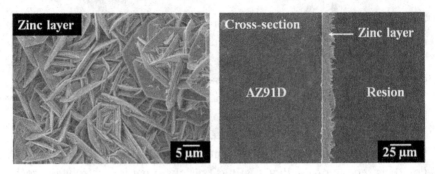

FIGURE 4.2 SEM micrograph of electrodeposited zinc surface and the cross-section view.

**Source: Progress in Organic Coatings (Elsevier).**

AZ91D (Mg-Al-Zn) is not easy. Directly electroplating the polymer on the Mg alloy may lead to the metal dissolution before the polymer deposits on the surface. In a directly oxidized magnesium alloy in a tartrate solution containing a Py monomer, the potential could not rise high enough to electropolymerize the Py monomer but remained in the active dissolution potential region of Mg. If the zinc layer was introduced between the Mg alloy and PPy layer, the zinc layer on the magnesium alloy could largely prevent the dissolution and enables the PPy to be electropolymerized on the surface. With this method, the PPy layer could be electropolymerized on the zinc-plated AZ91D in an aqueous sodium tartrate solution.

Figure 4.2 shows the SEM photographs of the surface and cross section of electroplated zinc surface. The compact zinc layer with uniform thickness and good adhesion was formed on the AZ91D alloy surface. The flake-like zinc crystal was able to provide enough adhesion between the inner AZ91D and outer PPy layer.

The PPy layer was formed by using galvanostatic electropolymerization method on the AZ91D alloy covered by zinc electroplating. Figure 4.3 shows the potential transient as a function of electric charge during galvanostatic electropolymerization of PPy at current densities (CD) from 5 to 20 $mA \cdot cm^{-2}$. And the duration of galvanostatic oxidation was 3,600 s at 5 $mA \cdot cm^{-2}$ and 900 s at 10, 15, and 20 $mA \cdot cm^{-2}$, respectively. The measured potential initially exhibited a plateau. The potential initially exhibited a plateau at the relatively low potential, which depended on a CD, then sharply rose to a peak, and finally gradually decreased to a constant value of. The potentials of the plateau and peak were increased

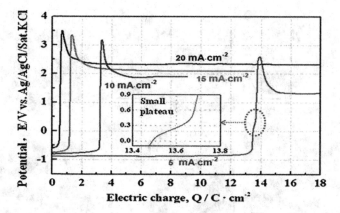

FIGURE 4.3 Potential transient during constant current oxidation at 5, 10, 15, and 20 mA·cm$^{-2}$ in 0.2 M sodium tartrate solution containing 0.5 M Py monomer as a function of electric charge applied.

**Source: Progress in Organic Coatings (Elsevier).**

with the increase of a CD and the electric charge to the potential peak greatly depended on the CD. The small plateau in the potential transient was observed in Figure 4.3, depending on CD at 0.3 V by 5 mA·cm$^{-2}$ and at 1.8 V by 20 mA·cm$^{-2}$, if the plot was widened in a horizontal direction. It is assumed that the zinc oxide is formed during the initial plateau and the zinc tartrate is formed around the small plateau. The formation of the zinc compounds in the initial stage will be discussed later.

Nucleation of PPy is started on the zinc compound oxide at around the peak potential. After the peak, a black PPy layer was observed. After the nucleation, the potential gradually fell down and was kept constant. In this period, the PPy layer continuously grew on the whole surface.

The surface morphology of a PPy layer on the AZ91D alloy covered by zinc electroplating was affected by the applied current density. As the FE-SEM observed (Figure 4.4), the PPy layer formed at 20 and 15 mA·cm$^{-2}$ exhibited a cauliflower-like appearance in which small spherical grains with a few μm diameter conglomerated. For the PPy layer formed at 10 mA·cm$^{-2}$, however, small spherical grains with 1 μm diameter were homogeneously distributed and the conglomerate feature was not observed. The size of the cauliflower-like conglomerate was a function of CD and, however, the size of the small spherical grain did not depend on CD.

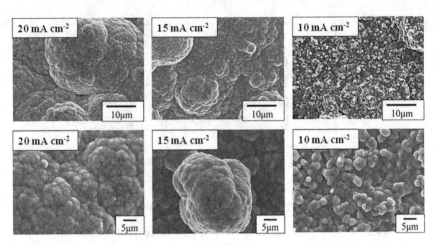

FIGURE 4.4    FE-SEM micrograph of PPy layer formed at 20 mA·cm$^{-2}$, 15 mA·cm$^{-2}$, and 10 mA·cm$^{-2}$ on two different scales.

**Source: created by author for this publication.**

## 4.4.2 The Effect of the Current Density and Molybdate Addition

Figure 4.5 shows the potential transient as a function of electric charge during two-step galvanostatic electropolymerization of PPy on the AZ91D alloy covered by zinc electroplating, and thus formed surface morphology, respectively. Figure 4.5 shows the potential transient during the formation of a PPy film in a solution of 0.2 M sodium tartart and 5 mM sodium molybdate, during the two-step galvanostatic electropolymerization, and thus formed surface morphology, respectively. During the two-step current-controlled synthesis, first we controlled the current at 15 mA cm$^{-2}$, and then at a lower CD of 1–5 mA cm$^{-2}$ for 3 C cm$^{-2}$. The total electric charge density for the PPy film formation was controlled at 4.5 C cm$^{-2}$. We prepared a PPy-Tart layer (Figure 4.5) and a PPy-Tart-Mo layer (Figure 4.6) to consider the effect of the current density during the galvanostatic electropolymeriazation of PPy. From the two potential transient figures showing similar tendency, a lower current density shows a lower potential and more compact surface morphology, although, PPy-Tart-Mo shows a lower the potential transient, it is possibly because of the catalytic effect of molybdate ions on PPy polymerization. The charge of initial plateau prior to the potential peak is larger in the presence of the molybdate ion than in its absence.

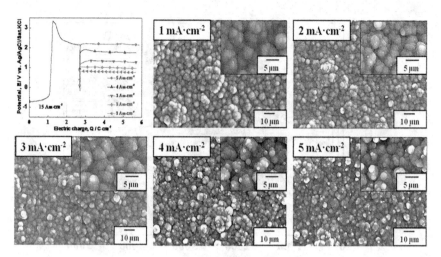

FIGURE 4.5    FE-SEM micrograph of the PPy layer formed at a mixed CD: first at 15 mA·cm$^{-2}$ for 1.5 C·cm$^{-2}$ and then changed to a lower CD of 1 to 5 mA·cm$^{-2}$ for 3 C·cm$^{-2}$ in 0.2 M sodium tartrate solution containing a 0.5 M Py monomer.

**Source: Progress in Organic Coatings (Elsevier).**

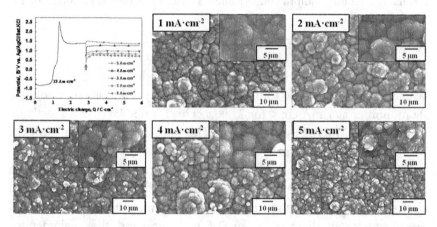

FIGURE 4.6    FE-SEM micrograph of PPy layer formed at a mixed CD: first at 15 mA·cm$^{-2}$ for 1.5 C·cm$^{-2}$ and then changed to a lower CD of 1 to 5 mA·cm$^{-2}$ for 3 C·cm$^{-2}$ in a 0.2 M sodium tartrate solution containing 0.5 M Py monomer.

**Source: Progress in Organic Coatings (Elsevier).**

## 4.4.3 The Effect of Ultrasonic Irradiation and Formation of PPy-Tart-MoO$_4$/ PPy-DS Bi-Layer

Figure 4.7 shows the PPy film formed in a sodium tartrate solution containing sodium molybdate initially at a CD of 15 mA·cm$^{-2}$ and then

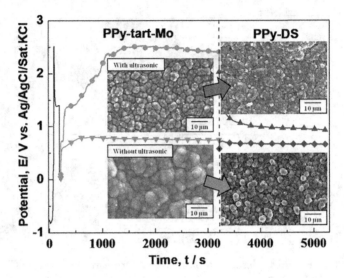

FIGURE 4.7 Galvanostatic electropolymerization of a bilayered PPy film of the inner layer in a 0.2 M sodium tartrate and 5 mM sodium molybdate solution was performed by the two-step galvanostatic oxidation and the outer layer formed in a 25 mM sodium dodecylsulfate solution potential-time curve and the SEM views.

**Source: Progress in Organic Coatings (Elsevier).**

at 1 mA·cm$^{-2}$ with or without ultrasonic irradiation; after that, the outer layer was formed in a sodium dodecylsulfate solution containing a Py monomer at 1 mA·cm$^{-2}$ for 2,000 s. This formed layer was shorthand for PPy-Tart-MoO$_4$/PPy-DS bilayer. The total charge used to form the PPy bilayer was 6.5 C·cm$^{-2}$.

The SEM images shown under ultrasonic irradiation of the PPy film (PPy-Tart-MoO$_4$) consists of the finer agglomerated cylindroids: the film electropolymerized without ultrasonic irradiation the small spherical protuberances in diameter from 0.4 to 2 μm and the agglomerated cylindroids are about 8.5 μm; and electropolymerized under ultrasonic irradiation the small spherical protuberances in diameter from 0.2 to 1 μm form and the agglomerated cylindroids are about 4 μm. This indicated that ultrasonic irradiation acted on the electropolymerization process, and thus formed a PPy layer that shows more compact and smooth. The outer layer formed on a PPy-Tart-MoO$_4$ layer was also affected by the ultrasonic irradiation step. When the inner layer was treated with ultrasonic irradiation, the outer PPy-DS layer exhibited more compactness and regularrity.

## 4.5 THE MECHANISM OF PPY FILM GROWTH ON A ZINC-COATED AZ91D ALLOY DURING GALVANOSTATIC ELECTROPOLYMERIZATION

In order to make the formation process clear, we try to examine the electropolymerization process from the observation of surface morphology by scanning electron microscopy (SEM) and the molecular vibration spectra by Raman spectroscopy.

The substrate is a zinc-coated AZ91D alloy. Electropolymerization of Py was carried out by constant current polarization at room temperature in an aqueous solution of 0.2 mol·dm$^{-3}$ (M) sodium tartrate containing a 0.5 M Py monomer. To avoid oxidation, the Py monomer was added into the solution by a syringe after the electrolyte was deoxygenated by bubbling nitrogen gas. The electropolymerization was carried out in a glass cell (40 cm$^{-3}$), which was equipped with a standard three-electrode system. An Ag/AgCl-saturated KCl electrode was used for the reference electrode and a platinum foil for the counter electrode. The reference electrode was connected via a salt bridge and a Luggin capillary. For the electrochemical measurement, a Hokuto-Denko HA-501 potentiostatic/galvanostatic was used and the date were recorded by a midi-logger, Graphtec GL200A. A field emission scanning electron microscope (FE-SEM, JEOL JSM-6500F) was used to observe the surface morphology. A Raman spectrometer (Bunko-Keiki model BRM-300) was used, in which excitation was made by a 532.0 nm wavelength laser light from a YVO$_4$ laser. The intensity of the laser light with 100 mW output was declined to about 5 mW by the neutral density filters to avoid destruction of the black PPy film.

The potentials of the initial plateau, the peak, and the last plateau increased with the increase of a CD (Figure 4.8). The electric charge during the initial plateau to the peak greatly depended on the CD, increasing with a decrease of CD. A black PPy film chould be observed after the peak in the potential-time curve. Nucleation of the PPy formation may start at the peak potential. After the nucleation, the potential gradually felt down and was kept constant. In this period, the PPy layer continuously grew on the whole surface.

To clarify the process in the plateau before the PPy polymerization, the potential change was plotted during the constant current oxidation at CD of 10 mA·cm$^{-2}$ as a function of time was studied. Raman spectra of the surface were measured at five points from (A) to (F), as shown in Figure 4.8. The Raman spectra of the five points as shown in our paper.[17] The assignment of the Raman bands listed in Table 4.1.

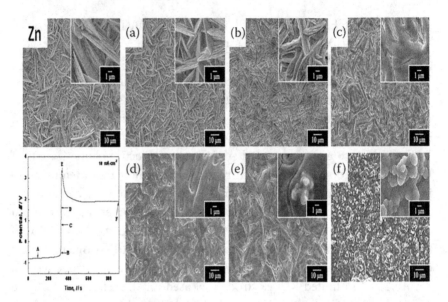

FIGURE 4.8   The galvanostatic electropolymerization of the PPy film at a CD of 10 mA·cm⁻² in a 0.2 M aqueous sodium tartarte containing a 0.5 M Py monomer, formed at points (a) −0.8 V, (b) −0.5 V, (c) 0.8 V, (d) 1.6 V, (e) 3.2 V, and (f) PPy film electropolymerized for 900s, and the corresponding FE-SEM micrograph of the surface of zinc electroplating.

**Source: created by author for this publication.**

The Raman spectra data shows that when the potential reached point (A) at −0.80 V, which is only 0.02 V higher than the open circuit potential before the imposition of 10 mA·cm⁻², a strong and broad Raman band appeared at 400–600 cm⁻¹. The peak at 565 cm⁻¹ is assigned to the ZnO and the weak peak at 322 cm⁻¹ may be assigned to Zn(OH)₂.[37–39] This pointed out that the zinc electroplating surface is covered with the ZnO and Zn(OH)₂ layer was detected. The Raman spectra of points (B) and (C) show three new bands between 800–1,100 cm⁻¹ and two weak bands between 1,000–1,200 cm⁻¹. These are in agreement with the Raman band of tartrate salts.[21,22] In the higher wave-number region, two peaks are seen at about 1,420 and 1,340 cm⁻¹, which were assigned to O-C=O symmetric stretching and C-H bending modes in tartrate, respectively, although the relative intensity of the higher wave-number Raman peaks is too large compared with the Raman peaks of sodium tartrate shown in the latter section. The sharp peaks at 1,560 cm⁻¹ and 322 cm⁻¹ are assumed to be assigned to those of a Py oligomer adsorbed on the surface compound.

TABLE 4.1 Raman Shifts and Assignment of Raman Bands for Solid Sodium Tartrate and for the PPy Layer fOrmed in a 0.2 M Aqueous Sodium Tartrate Containing a 0.5 M Py Monomer at a CD of 10 mA·cm$^{-2}$ When the Potential Reaches Values of −0.8 V, −0.5 V, and 0.8 V

| Experimental Raman shift, Δν/cm$^{-1}$ | | | | Assignments |
|---|---|---|---|---|
| Sodium tartrate | The point potential, E/V | | | |
| | (a) −0.8 V | (b) −0.5 V | (c) 0.8 V | |
| 1606 | 1606 | 1635 | 1649 | O-C=O antysymmetric stretching mode |
| | 1560 | 1560 | 1560 | Py oligomer C=C stretching |
| 1421 | 1445 | 1425 | 1416 | O-C=O symmetric stretching mode |
| 1312 | 1342 | 1350 | | C-H bending mode |
| 1232 | | 1224 | 1229 | |
| 1120 | | 1110 | 1110 | C-OH stretching mode |
| 1069 | | 1051 | 1051 | |
| 997 | | 1007 | 1007 | C-C stretching mode |
| 897 | | 929 | 929 | |
| 845 | | | | Deformation mode of O-C=O |
| 808 | | 812 | 815 | groups |
| 709 | | | | |
| | | 656 | 656 | |
| 613 | | | | |
| | 565 | 557 | 556 | ZnO |
| 529 | | | | Deformation mode of O-C=O groups |
| | 494 | 494 | 495 | ZnO |
| 489 | | | | Deformation mode of O-C=O groups |
| 372 | | | | Tartrate skeletal deformation mode |
| | 322 | 322 | 322 | Zn(OH)$_2$/Py oligomer C=C stretching |

Source: Elsevier.

It is seen that the intensity of the Raman band for zinc oxide decreases from −0.8 V to 0.8 V, while the intensities of Raman bands for zinc tartrate and Py oligomer increase.

The Raman spectra of the surface taken at points (D), (E), and (F) clearly are in accordance with PPy peaks,[30,40–42] and the intensity of the peaks increases with the time duration from point (D) to point (F). There were no Raman bands corresponding to the tartrate anion doped. The assignments of the Raman bands are listed in Table 4.2.

TABLE 4.2   Raman Shifts and Assignment of Raman Bands for the PPy Layer Formed in a 0.2 M Aqueous Sodium Tartrate Containing a 0.5 M Py Monomer at a CD of 10 mA·cm$^{-2}$ When the Potential Reaches Values of 1.6 V and 3.2 V, Before and After Electropolymerization

| The point potential/V Wavenumbers/cm$^{-1}$ | | | Assignments |
|---|---|---|---|
| (d) 1.6 V | (e) 3.2 V | (f) PPy film | |
| 1576 | 1590 | 1581 | conjugative C=C stretching |
| 1419 | 1429 | 1414 | C-N stretching |
| 1328 | 1327 | 1328 | C-N stretching or ring stretching of PPy |
|  | 1259 | 1258 | antisymmetrical C-H in-plane deformation |
| 1045 | 1046 | 1048 | symmetrical C-H in-plane deformation |
| 989 | 988 | 990 | ring deformation |
| 933 | 923 | 934 | C-H out of the plane deformation |

*Source:* Elsevier.

The morphology of the surface from points (A) to (F) was also observed by SEM to intuitive analysis the mechanism. The surface views are shown in Figure 4.8, in which two different magnification images are shown. The initial surface of the zinc-coated AZ91D alloy before the oxidation is also presented as a contrast.

On the surface of the zinc-coated alloy, zinc crystal plates stood diagonally to the surface. At point (A) in the initial stage of the plateau, the surface morphology did not largely change. The edge of the zinc crystal plate was gradually indistinct at points (B) and (C). At points (D) and (E), the zinc crystal plates disappeared and the smooth and homogeneous surface appeared. At point (E), there were small spherical particles with a few nm diameter dotted on the surface. Finally, the surface was completely covered with the spherical particles at point (F).

In summary, the electropolymerization processes estimated from the above Raman spectra and the SEM views are schematically presented in Figure 4.9. The formation of PPy by galvanostatic electropolymerization is strongly influenced by the electrode material and the electrolyte.[43,44] The dissolution of the working electrode of the magnesium alloy is prevented by a zinc electroplating layer on the AZ91D alloy surface.

Firstly, as the potential arrived at point (A), the plated zinc crystals dissolved with the simultaneous formation of the ZnO and $Zn(OH)_2$ on their surface. Then potential arrived at point (B) at the end in the plateau

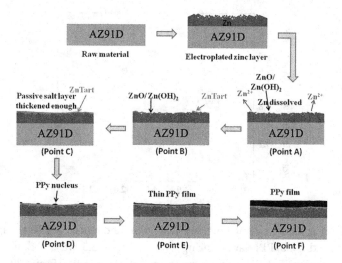

FIGURE 4.9 Schematic of the galvanostatic electropolymerization process of a PPy film growth on an AZ91D alloy with an electroplated zinc layer.

**Source: created by author for this publication.**

layer of ZnO and $Zn(OH)_2$ was gradually replaced with a zinc-tartrate salt layer, according to the reactions:

$$ZnO + 4H^+ \rightarrow Zn^{2+} + 2H_2O \tag{4.1}$$

$$Zn^{2+} + C_4H_4O_6{}^{2-} \rightarrow ZnC_4H_4O_6 \tag{4.2}$$

After the zinc salt layer was thickened enough to protect the zinc as well as the inner magnesium alloy (around point C), PPy nuclei began to form on the surface (around point D). Then, at the peak potential at point (E), the surface was almost covered by the PPy nuclei. Finally, when the potential gradually fell down to the second plateau, the black and homogeneous PPy film grew three-dimensionally and covered the whole surface.

## 4.6 THE CORROSION PERFORMANCE OF THE KINDS OF PPY LAYERS

The AZ91D alloy covered by zinc and the PPy thus electropolymerized was immersed in the 3.5 wt.% sodium chloride solution at 25°C in order to inspect the corrosion protection of the PPy coating. During the

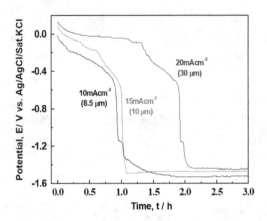

FIGURE 4.10 Open circuit potential (OCP) of zinc-coated AZ91D magnesium alloy covered by a PPy layer during exposure in a 3.5 wt.% sodium chloride aqueous solution. The PPy layers were formed at a CD of 10, 15, and 20 mA·cm$^{-2}$ for 900 s in a 0.2 M sodium tartrate solution containing a 0.5 M Py monomer.

**Source: Progress in Organic Coatings (Elsevier).**

immersion, the surface of the alloy was visually observed for the rust generation. The transient of open circuit potential (OCP) during the immersion exhibited the corrosion performance of the PPy layers, respectively.

Figure 4.10 shows the OCP of PPy layers formed on zinc-coated AZ91D magnesium alloy by constant current control with 10, 15, and 20 mA·cm$^{-2}$ for 900 s and the thickness of the PPy layer was estimated to be about 8.5, 10, and 30 μm from the cross-section SEM view, respectively.

When the potential was higher than −1.0 V, no corrosion products were observed. When the potential decayed lower than −1.0 V, white corrosion products immediately appeared. Thus, the time period before the potential decays to −1.0 V is assumed to be a protection time, in which the PPy-coated substrate was possibly kept at the passive state. The protection time found from Figure 2.8 greatly depends on the thickness of the PPy layer, increasing with the thicker PPy layer covering the alloy.

Figure 4.11 shows the OCP of an AZ91D alloy covered by zinc electroplating coated by a PPy-Tart layer formed by two-step electropolymerization. First, a CD of 15 mA·cm$^{-2}$ for 100 s and then 1, 2, 3, 4, or 5 mA·cm$^{-2}$ with a total electricity of 4.5 C·cm$^{-2}$ (Figure 4.5). The coated

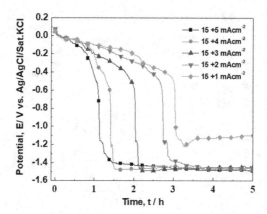

FIGURE 4.11 Open circuit potential of zinc-coated AZ91D magnesium alloy covered by the PPy layer during exposure in a 3.5 wt.% sodium chloride aqueous solution. The PPy layers were formed at a mixed CD of 15 mA·cm$^{-2}$ for 100 s and then changed to a lower CD for 3 C·cm$^{-2}$ in a 0.2 M sodium tartrate solution containing a 0.5 M Py monomer.

**Source: Progress in Organic Coatings (Elsevier).**

AZ91D alloy initially exhibited high potential values from 0.08 to −0.5 V, which indicates that the AZ91D alloy was passivated by the oxidative PPy layer, then the potential gradually was reduced to an active state potential at about −1.4 V. When the potential was more positive than −0.9 V, no corrosion products were observed on the black PPy layer. When the OCP reached approximately −1.2 V in the active state, however, white corrosion product was clearly observed on the surface. The protection time is thus determined by the time length until the potential decayed to −0.9 V. Therefore, we found that the PPy layer formed by the lower CD in the second oxidation of the two-step polymerization protected AZ91D more effectively against corrosion.

Figure 4.12 shows the effect of a molybdate anion incorporated in the PPy film on the protection of AZ91D alloy against corrosion. The PPy-Tart-MoO$_4$ layer was formed by two-step electropolymerization. First, a CD of 15 mA·cm$^{-2}$ for 100 s and then at 1 mA·cm$^{-2}$ for 3,000 s are imposed. The layers were prepared in the taryrate solution containing sodium molybdate at different concentrations (1, 3, 5, 7, and 10 mM). The coated AZ91D alloy exhibits potential at about 0 V in the passive region during the initial immersion from 7 to 13 hours depending on molybdate concentration for PPy preparation. The potential then

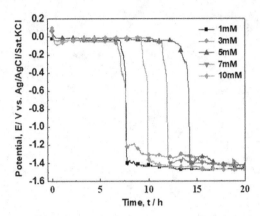

FIGURE 4.12 Open circuit potential of zinc-coated AZ91D magnesium alloy covered by a PPy layer during exposure in a 3.5 wt.% sodium chloride aqueous solution. The PPy layers were formed at a mixed CD of 15 mA·cm$^{-2}$ for 100 s and then changed to 1 mA·cm$^{-2}$ for 3,000 s in sodium tartrate and different concentration disodium molybdate solution containing a 0.5 M Py monomer.

**Source: Progress in Organic Coatings (Elsevier).**

gradually decreased to the active potential region of an AZ91D alloy. The addition of molybdate anion is seen to prolong the corrosion protection time by the PPy layer and, however, the optimal concentration of molybdate added is 5 mM. Because, when the concentration of molybdate is higher than 5 mM, the protection time decreased.

Figure 4.13 shows the OCP of the bilayered coating of PPy-Tart-MoO$_4$/PPy-DS prepared with or without ultrasonic irradiation. The AZ91D alloy covered by a PPy film prepared without ultrasonic irradiation initially exhibits potential at 0.08 V and then gradually falls down to stable potential at around 0.04 V and kept for 72 hours of immersion. After 72 hours, the potential decreases to the active potential region of AZ91D and white corrosion products could be observed on surface. The AZ91D alloy covered by the bilayered film, which an inner PPy layer was electropolymerized under ultrasonic irradiation, exhibited better performance. The OCP curve shown initially presents a potential at 0.29 V and gradually decreased to a stable potential at 0.04 V. The potential was kept for 221 hours in which no appreciable corrosion took place. During the period, the potential decreased several times to around −0.2 V, and then returned to 0.04 V. After 221 hours, it gradually decreased to the active potential region of AZ91D.

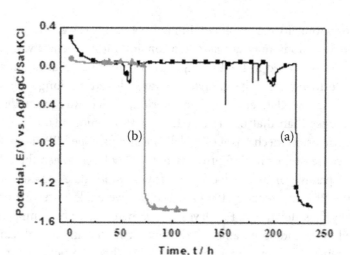

FIGURE 4.13   Open circuit potential (OCP) of zinc-coated AZ91D magnesium alloy covered by a bilayered PPy layer during exposure in a 3.5 wt.% sodium chloride aqueous solution. The inner layer was formed at a mixed CD and with (a) or without (b) ultrasonic irradiation in a sodium tartrate and disodium molybdate solution and the outer layer was formed in a sodium dodecylsulfate solution.

**Source: Progress in Organic Coatings (Elsevier).**

Therefore, we can conclude that the corrosion-resistant performance of the PPy layer is affected by both the flatness of the surface and also the addition of the positive ion. By using a single CD, the PPy film could only be kept at a passive state only for 1 hour. In addition, by using two-step CDs, the PPy film showed the better corrosion performance. The PPy layer electropolymerized at a lower CD can protect the AZ91D alloy for a longer time period because of the more compact and denser morphology PPy surface. The protective property is assumed to be due to the physical property of the PPy layer, the dense PPy layer acts as a better barrier which prevents chloride ions from penetrating through the PPy film into the passive layer. When chloride ions contact the inner passivating layer, the ions break the passive layer, causing the immersion potential to be decreased to the active potential region.

On the other hand, when molybdate present in the tartrate solution during the PPy formation, protection property of the PPy layer against corrosion of AD91D was enhanced. The PPy-Tart-$MoO_4$ layer thus electropolymerized by the two-step CD scan keeps the alloy in a passive state for a longer time, as shown in Figure 4.6. It was well known that the molybdate anions may act as corrosion inhibitors themselves. $MoO_4^{2-}$ ions could be stored in the PPy layer as dopant ions, resulting in a much better performance. The molybdate ions doped prolonged inhibitive property for a relatively long time period, which was more than four times longer than that of the PPy layer without molybdate.

The corrosion mechanism of molybdate anions doped PPy layer was not only because of the physical property of the PPy layer. When the PPy layer and the passive oxide are locally destroyed, molybdate ions are released from the PPy and repaired the damaged passive oxide. After the potential gradually falls down, the PPy layer is assumed to be not completely in a reduced state and some amount of doped counter anions still exist in a polymer matrix in the potentials to −0.5 V. Then the potential gradually decreased during further immersion to the active potential region of an AZ91D alloy in which the PPy is in reduced state and lost the protection ability against corrosion of the AZ91D alloy.

The AZ91D alloy covered by a bilayered PPy film of PPy-Tart-MoO₄/ PPy-DS maintained a passive state longer than the PPy-Tart or PPy-Tart-MoO₄ layer in a sodium chloride solution. Since the total electric charge for electropolymerization of the bilayered PPy was 6.5 C·cm$^{-2}$, it was by 2 C·cm$^{-2}$ larger than that of a single PPy layer. The bilayered structure prolonged the corrosion protection time to 75 hours, which was 60 hours longer than that of a single PPy layer. The DS ions doped play an important role, though the thicker PPy layer can protect an AZ91D alloy for a longer time.

The corrosion protection mechanism of the bilayered PPy film could be explained in the following ways. The DS doping ions with relatively large size have lower mobility than the molybdate anions from the inner layer; thus, the outer PPy-DS layer inhibits the release of molybdate anions from the inner PPy layer to an aqueous solution. In other words, it is assumed that molybdate anions remain in the inner PPy layer for a longer time and the stabilized passive oxide film on the AZ91D alloy covered zinc electroplating more than the single PPy layer.

Further, when the inner PPy-Tart-MoO₄ layer was prepared under imposition of ultrasonic irradiation and the layer was covered by the outer PPy-DS layer, the inhibition property is much improved. The surface of the bilayered PPy film formed under the ultrasonic irradiation consisted of finer particles and revealed smoother. The molybdate ions doped in the PPy layer may be a larger amount than that of the bilayered

PPy layer prepared without ultrasonic irradiation. In addition, the more compact and dense PPy-Tart-MoO$_4$ layer may stabilize the passive oxide film for a longer period of time. Therefore, the bilayered PPy film with the inner layer formed under imposition of an ultrasonic wave improved structural properties and can induce a large duration of passivation during immersion in the sodium chloride solution.

## 4.7 SUMMARY

For corrosion protection, three different types of PPy films formed on the AZ91D alloy covered by zinc electroplating, PPy-Tart film, PPy-Tart-MoO$_4$ film, and the bilayered PPy film of PPy-Tart-MoO$_4$/ PPy-DS. A dense PPy layer was formed by using two-step constant-current electropolymerization, which induced the passivation of the zinc-coated AZ91D in an aqueous 3.5 wt.% sodium chloride solution and the passivation continued for several hours. Doping of molybdate ions into the PPy-Tart layer significantly improved the protective property of the film. The longest duration of a PPy-Tart-MoO$_4$ single layer formed by two-step constant-current electropolymerization prolonged the passivation for 14.2 hours during immersion in the sodium chloride solution. The bilayered PPy coating consisting of an inner PPy-Tart-MoO$_4$ and outer PPy-DS layer kept the zinc-coated AZ91D alloy passive for 72 hours. The better corrosion protection may originate in the large size of DS ions that restrict the release of MoO$_4$ anions from the inner layer. Ultrasonic irradiation during the electropolymerization of the inner PPy-Tart-MoO$_4$ layer further improved the protection property of the bilayered PPy of PPy-Tart-MoO$_4$/PPy-DS against corrosion of the zinc-coated AZ91D. The time maintaining the passive state of the alloy reached 221 hours.

## REFERENCES

1. Fekry, A. M. & Fatayerji, M. Z. 2009. Electrochemical corrosion behavior of AZ91D alloy in ethylene glycol. *Electrochimica Acta*, **54**, 6522–6528.
2. Toros, S., Ozturk, F. & Kacar, I. 2008. Review of warm forming of aluminum-magnesium alloys. *Journal of Materials Processing Technology*, **207**, 1–12.
3. Gastaldi, D., Sassi, V., Petrini, L., Vedani, M., Trasatti, S. & Migliavacca, F. 2011. Continuum damage model for bioresorbable magnesium alloy devices - Application to coronary stents. *Journal of The Mechanical Behavior of Biomedical Materials*, **4**, 352–365.

4. Pollock, T. M. 2010. Weight loss with magnesium alloys. *Science*, **328**, 986–987.

5. Jonsson, M., Persson, D. & Leygraf, C. 2008. Atmospheric corrosion of field-exposed magnesium alloy AZ91D. *Corrosion Science*, **50**, 1406–1413.

6. Eliezer, A., Gutman, E. M., Abramov, E. & Unigovski, Y. 2001. Corrosion fatigue of die-cast and extruded magnesium alloys. *Journal of Light Metals*, **1**, 179–186.

7. Kulekci, M. K. 2008. Magnesium and its alloys applications in automotive industry. *The International Journal of Advanced Manufacturing Technology*, **39**, 851–865.

8. Ocon, P., Cristobal, A. B., Herrasti, P. & Fatas, E. 2005. Corrosion performance of conducting polymer coatings applied on mild steel. *Corrosion Science*, **47**, 649–662.

9. Rohwerder, M. & Michalik, A. 2008. Conducting polymers for corrosion protection: What makes the difference between failure and success? *Electrochimica Acta*, **53**, 1300–1313.

10. Armelin, E., Oliver, R., Liesa, F., Iribarren, J. I., Estrany, F. & Alemán, C. 2007. Marine paint fomulations: Conducting polymers as anticorrosive additives. *Progress in Organic Coatings*, **59**, 46–52.

11. Sabouri, M., Shahrabi, T., Farid, H. R. & Hosseini, M. G. 2009. Polypyrrole and polypyrrole-tungstate electropolymerization coatings on carbon steel and evaluating their corrosion protection performance via electrochemical impedance spectroscopy. *Progress in Organic Coatings*, **64**, 429–434.

12. Martins, J. I., Costa, S. C., Bazzaoui, M., Gonçalves, G., Fortunato, E. & Martins, R. 2006. Conditions to prepare PPy/Al$_2$O$_3$/Al used as a solid-state capacitor from aqueous malic solutions. *Journal of Power Sources*, **160**, 1471–1479.

13. Kowalski, D., Ueda, M. & Ohtsuka, T. 2007. Corrosion protection of steel by bi-layered polypyrrole doped with molybdophosphate and naphthalenedisulfonate anions. *Corrosion Science*, **49**, 1635–1644.

14. Kowalski, D., Ueda, M. & Ohtsuka, T. 2007. The effect of counter anions on corrosion resistance of steel covered by bi-layered polypyrrole film. *Corrosion Science*, **49**, 3442–3452.

15. Kowalski, D., Ueda, M. & Ohtsuka, T. 2008. The effect of ultrasonic irradiation during electropolymerization of polypyrrole on corrosion prevention of the coated steel. *Corrosion Science*, **50**, 286–291.

16. Sheng, N. & Ohtsuka, T. 2012. Preparation of conducting poly-pyrrole layer on zinc coated Mg alloy of AZ91D for corrosion protection. *Progress in Organic Coatings*, **75**, 59–64.

17. Sheng, N., Ueda, M. & Ohtsuka, T. 2013. The formation of polypyrrole film on zinc-coated AZ91D alloy under constant current characterized by Raman spectroscopy. *Progress in Organic Coatings*, **76**, 328–334.

18. Sheng, N., Lei, Y. H., Hyonoo, A., Ueda, M. & Ohtsuka, T. 2014. Improvement of polypyrrole films for corrosion protection of zinc-coated AZ91D alloy. *Progress in Organic Coatings*, **77**, 1724–1734.

19. Esmaily, M., Svensson, J. E. & Fajardo, S. 2017. Fundamentals and advances in magnesium alloy corrosion. *Progress in Materials Science*, **89**, 92–193.

20. Bussy, A. 1830. Chlorure De Magnésium. *Journal de chimie médicale de pharmacie et de toxicologie*, 141.

21. Beetz, W. 1866. On the development of hydrogen from the anode. *The London, Edinburgh, and Dublin Philosophical Magazine and Journal of Science*, **32**, 269–278.

22. Ali, M., Hussein, M. A. & Al-Aqeeli, N. 2019. Magnesium-based composites and alloys for medical applications: A review of mechanical and corrosion properties. *J. Alloy. Compd.*, **792**, 1162–1190.

23. Kirkland, N. T., Birbilis, N. & Staiger, M. P. 2012. Assessing the corrosion of biodegradable magnesium implants: A critical review of current methodologies and their limitations. *Acta Biomater.*, **8**, 925–936.

24. Erbel, R., Di Mario, C. & Bartunek, J. 2007. Temporary scaffolding of coronary arteries with bioabsorbable magnesium stents: a prospective, non-randomised multicentre trial. *Lancet*, **369**, 1869–1875.

25. Ding, W. J. 2016. Opportunities and challenges for the biodegradable magnesium alloys as next-generation biomaterials. *Regenerative biomaterials.*, **3**, 79–86.

26. Heublein, B., Rohde, R., Kaese, V., Niemeyer, M., Hartung, W. & Haverich, A. 2003. Biocorrosion of magnesium alloys: a new principle in cardiovascular implant technology? *Heart*, **89**, 651–656.

27. Xu, D. K., Han, E. H. & Xu, Y. B. 2016. Effect of long-period stacking ordered phase on microstructure, mechanical property and corrosion resistance of Mg alloys: A review. *Progress in Natural Science: Materials International.*, **26**, 117–128.

28. Guimard, N. K., Gomez, N. & Schmidt, C. E. 2007. Conducting polymers in biomedical engineering. *Progress in Polymer Science*, **32**, 876–921.

29. Mohammad, F. 2001. Stability of electrically conducting polymers - ScienceDirect. *Handbook of Advanced Electronic & Photonic Materials & Devices*, Academic Press, **8**, 321–350.

30. Bazzaoui, M., Martins, L., Bazzaoui, E. A. & Martins, J. I. 2002. New electrochemical procedure for elaborating homogeneous and strongly adherent PPy films on zinc electrodes. *Journal of Electroanalytical Chemistry*, **537**, 47–57.

31. Martins, J. I., Reis, T. C., Ba Zzaoui, M., Ba Zzaoui, E. A. & Martins, L. 2004. Polypyrrole coatings as a treatment for zinc-coated steel surfaces against corrosion. *Corrosion Science*, **46**, 2361–2381.

32. Dos Santos, L. M., Lacroix, J. C., Chane-Ching, K. I., Adenier, A., Abrantes, L. M. & Lacaze, P. C. 2006. Electrochemical synthesis of

polypyrrole films on copper electrodes in acidic and neutral aqueous media. *Journal of Electroanalytical Chemistry*, **587**, 67–78.

33. Lenz, D. M., Delamar, M. & Ferreira, C. A. 2007. Improvement of the anticorrosion properties of polypyrrole by zinc phosphate pigment incorporation. *Progress in Organic Coatings*, **58**, 64–69.

34. Ma, R., Sask, K. N., Shi, C., Brash, J. L. & Zhitomirsky, I. 2011. Electrodeposition of polypyrrole-heparin and polypyrrole-hydroxyapatite films. *Materials Letters*, **65**, 681–684.

35. Martins, J. I., Costa, S. C. & Bazzaoui, M. 2006. Conditions to prepare PPy/Al$_2$O$_3$/Al used as a solid-state capacitor from aqueous malic solutions. *Journal of Power Sources*, **160**, 1471–1479.

36. Martins, J. I., Bazzaoui, M., Reis, T. C., Costa, S. C., Nunes, M. C., Martins, L. & Bazzaoui, E. A. 2009. The effect of pH on the pyrrole electropolymerization on iron in malate aqueous solutions. *Progress in Organic Coatings*, **65**, 62–70.

37. Ohtsuka, T. & Matsuda, M. 2003. In Situ Raman Spectroscopy for Corrosion Products of Zinc in Humidified Atmosphere in the Presence of Sodium Chloride Precipitate. *Corrosion, The Journal of Science and Engineering*, **59**, 407–413.

38. Kasperrek, J. & Lenglet, M. 1997. Identification of thin films on zinc sucstrates by FTIR and Raman spectroscopies. *Revue de métallurgie.*, **94**, 713–719.

39. Hugot-Le Goff, A., Joiret, S., Saidani, B. & Wiart, R. 1989. In-situ Raman spectroscopy applied to the stidy of the deposition and passivation of zinc in alkaline electrolytes. *Journal of Electroanalytical Chemistry*, **263**, 127–135.

40. Duchet, J., Legras, R. & Demoustier-Champagne, S. 1998. Chemical synthesis of polypyrrole: structure-properties relationship. *Synthetic Metals*, **98**, 113–122.

41. Chen, F. E., Shi, G., Fu, M., Qu, L. & Hong, X. 2003. Raman spectroscopic evidence of thickness dependence of the doping level of electrochemically deposited polypyrrole film. *Synthetic Metals*, **132**, 125–132.

42. Santos, M. J. L., Brolo, A. G. & Girotto, E. M. 2007. Study of polaron and bipolaron states in polypyrrole by in situ Raman spectroelectrochemistry. *Electrochimica Acta*, **52**, 6141–6145.

43. Rizzi, M., Trueba, M. & Trasatti, S. P. 2011. Polypyrrole films on Al alloys: The role of structural changes on protection performance. *Synthetic Metals*, **161**, 23–31.

44. Nalwa, H. S. 2001. H. S. Nalwa, Ed., *Handbook of Advanced Electronic and Photonic Materials and Devices, 10-Volume Set*, Academic Press, San Diego.

# CHAPTER 5

# Applications of Conducting Polymers for Anti-Corrosion in Marine Environments

## Qing Chen and Yanhua Lei

CONTENTS

DOI: 10.1201/9781003376194-5

## 5.1 INTRODUCTION TO MARINE CORROSION

The ocean covers about 70% of Earth's surface area and ocean transport supports 90% of freight transportation in the world trade. Materials are subjected to seawater conditions in numerous applications; for example, ships, pleasure boats, submarines, offshore platforms, subsea pipelines and telecommunications cables, wharfs, seawater-cooled power and chemical plants, desalination plants, fishing gear, and so on. [1] Seawater is a critical resource for food (primarily fish), table salt, and conversion to freshwater (desalination). It is also the primary source for extraction of magnesium and subsequent production of its alloys. Thus, marine resources and the marine industry have become indispensable pillars in economic development.

### 5.1.1 Environmental Characteristics of Seawater

Seawater itself is a native excellent electrolyte with high corrosiveness. The ocean environment is complicated because marine organisms and their metabolites will influence the materials together to cause corrosion. Corrosion of materials is always the main reason to cause destruction and abandonment of infrastructure and industrial equipment served in the marine environment. It has been recognized all around the world that corrosion losses exceed the total loss of all other natural disasters.[2,3]

The high salt concentration in most marine environments combined with high electrical conductivity make marine bodies a highly conducive environment for corrosion to occur on metal surfaces. The corrosion behaviors are especially varied and affected by marine environmental factors such as hydrostatic pressure, seawater temperature, concentration of dissolved oxygen and Cl⁻, pH value, fouling of microorganisms, and flow of seawater[1]. Hence, it is sometimes described as a 'living' medium and considered to be the most corrosive of the natural environments.

## 5.1.2 Corrosion and Protection Mechanisms

In such an aggressive and intricacy environment, various metals or alloys usually suffer from serious dissolution, even for the excellent corrosion-resisting alloys such as Ti, stainless steel, and copper alloys. From a corrosion standpoint, chloride ions are considered to be the most aggressive constituent in seawater. The precise role of chloride in the corrosion process is still not fully understood. For instance, chloride contributes to increased electrolytic conductivity by ion transport. Higher conductivity means that current between anodic and cathodic areas can flow over larger distances, for example, in galvanic couples; the magnitude of local cell currents can also be higher. Thus, the overall effect is typically higher in general and/or localized corrosion rates. Corrosion attacks at anodic areas are supported by reduction reaction(s) at cathodic sites – typically oxygen reduction in aerated seawater. Oxygen reduction is obviously a critical reaction for metallic corrosion in seawater.

This corrosion progress of steel in the marine environment is shown in Figure 5.1. The interface between metal and seawater is the most important factor affecting metal corrosion, so making a layer of coating on such an interface can greatly delay metal corrosion.

Marine corrosion can be prevented through careful study and use of corrosion-reducing materials, designs, and coatings.

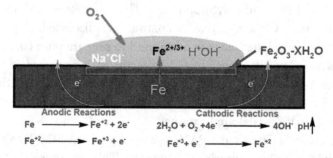

FIGURE 5.1 Schematic diagram of iron and steel corrosion.

Source: created by author for this publication.

At present, a coating is the most directly and widely used strategy in marine anti-corrosion, but with the demand of environmentally friendly protection, efforts have been made by researchers to seek efficient and environmentally friendly coatings. From this point of view, conductive polymers (CPs) have attracted extensive attention because of their excellent anti-corrosion properties.[4–8]

## 5.2 ANTI-CORROSION BASED ON CONDUCTING POLYMER (CP) COATINGS

Coatings on the surface of metals by polymeric materials have been widely used in industries for the protection of these materials against corrosion. In recent years, CPs have attracted great attention due to their stability, environmentally friendly nature, and other characteristics, making them suitable coating candidates for corrosion-resistant applications. The anti-corrosion performance of CPs can be traced back to 1985. Deberry et al.[9] found that the electrochemically deposited PANI coating shifted the positive corrosion potential of 410 stainless steel by 0.5 V (versus SCE), greatly reducing the corrosion rate. CPs have been intensively studied as effective additions into various commercial coatings for enhancing the anti-corrosion property of the coatings. This chapter mainly introduces the recent research progress of CP-blended coatings.

## 5.3 PURE CP COATINGS

CPs can prevent corrosion through a variety of mechanisms, which could be used in combination with most traditional coatings to achieve the desired anti-corrosion effect according to different needs in the industry. Typical CPs used in anti-corrosion coatings include polyaniline (PANI), polypyrrole (PPy), and polythiophene (PTh). At present, various types of commonly used CP anti-corrosive coatings are discussed.

Electrochemical methods are often used to prepare pure CPs coatings, which electrodeposit a layer of CPs directly on the surface of the substrate, such as PANI, PPy, PTh, and their derivates. These methods are suitable for small devices or laboratory tests, while they could not be used in large-scale industrial applications.

Mrad et al.[10] deposited a PANI coating on an AA2024-T3 aluminum alloy using an aniline sulfuric acid solution. The anti-corrosion property and mechanisms were analyzed and discussed. Wang et al.[11] first in-situ tested the electrodeposition process of aniline with a wire beam electrode

(WBE). With the experimental setup, they claimed that the WBE was a practical tool for monitoring, characterizing, and optimizing PANI electrodeposition processes, and PANI coatings were found to prevent localized corrosion of AA1100, primarily by enhancing its passive film rather than by a barrier mechanism.

In a pure PPy and PTh coating study, the method of deposition is also widely used for research. Arenas et al.[12] deposited a protective PPy coating on AA2024, and found that the partial reduction of the polymer structure due to the heat treatment made the PPy layer act as a physical barrier against corrosion. Kousik et al.[13] attempted to generate PTh coatings on a mild steel surface using acetonitrile as a medium, and determined that PTh-coated mild steel was protected by a passivation mechanism caused by the redox activity of the PTh. Water uptake and delaminating area studies also confirmed protective action of electro-polymerized PTh on the mild steel surface.

Other CPs synthesized and tested as anti-corrosive coatings include polyindole,[14] poly(3,4-ethylenedixythiophene),[15] poly(phenylenedia-mine),[16] poly(p-phenylene),[17], and poly(carbazole)[18] (Table 5.1).

## 5.4 CPS COMPOSITE BLENDED RESIN COATINGS

Traditional coatings usually include resin and fillers, which have the characteristics of simple construction and good anti-corrosive property. The PANI/resin composite coatings obtained by modification resins using PANI nanoparticles have advantages such as better corrosion resistance than traditional coatings, and have a greater development prospect than electrochemical deposition. At present, scientists have studied many kinds of resin coatings, such as PANI-modified epoxy resins,[19] polyvinyl bu-tyral,[20] polyurethane,[21] alkyd resins,[22], and acrylic resins,[23] of which PANI/epoxy coatings are the most widely studied. It's found that the corrosion resistance and service life of each resin modified with PANI were improved compared to pure resin coatings.

Besides PANI, other CPs, such as PPy, PTh, and PEDOT, were also employed to strengthen the protective property of various reign coat-ings.[24–26] Contri et al.[27] developed epoxy resin, epoxy/montmoril-lonite (Mt), epoxy/PPy, and epoxy/Mt-PPy-based coatings. According to EIS results in Figure 5.2, epoxy/Mt-PPy with a 5 wt.% content had fine corrosion protection for carbon steel. Consequently, a PPy-filled-epoxy system has been effectively used to prevent corrosion on various types of surfaces of steel.

TABLE 5.1  A Brief Description of Some Conducting Polymers Evaluated for Corrosion Protection

| S/N | Name | Abbreviation | Structure | Properties |
|---|---|---|---|---|
| 1 | Polyaniline | PANI | | PANI is a semi-flexible rod polymer that exists in three different forms: leucoemeraldine (white); emeraldine (green for the emeraldine salt, blue for the emeraldine base); and (per)nigraniline (blue/violet). PANI in emeraldine state is the most utilized due to its high stability at room temperature. The doped emeraldine salt exhibits high electrical conducting. Leucoemeraldine and pernigraniline are poor conductors. |
| 2 | Polypyrrole | PPy | | PPy is an amorphous, yellowish polymer that darkens in air due to oxidation. Oxidized form exhibits good conducting and electrical properties. The doped films are blue or black in color depending on the degree of polymerization and film thickness. PPy is insoluble in solvents but swellable. |
| 3 | Polyindole | PIN | | PIN is a polymer derived from the polymerization of indole, an aromatic bicyclic structured compound consisting of a six-membered benzene ring fused to a five-membered nitrogen-containing pyrrole ring. It has good blue photoluminescent characteristics, highly stable redox activity, fast switchable electrochromic ability. The doped form has good electrical conductivity and is stable in air. |

4    Polythiophene    PTh

PTh is a polymerized thiophene, a sulfur heterocycle compound. The polymer exhibits conducting and electrical conductivity when in oxidized form. PTh demonstrates interesting optical properties due to the conjugated backbone. The optical properties respond to environmental stimuli, with interesting colour shifts in response to changes in solvent, temperature, applied potential, and binding to other molecules.

5    Poly(phenylenediamine)    PPD

PPD is a ladder-type aromatic semiconductive polymer containing pyrazine and phenazine rings in its backbone. PPDs are synthesized by using o-, m-, and p-phenylenediamines.

*Source:* Elsevier.

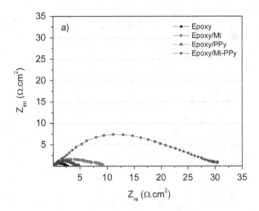

FIGURE 5.2   EIS graphs for steel-coated epoxy coating and respective composites coatings with 5 wt.% of Mt, PPy, and Mt-PPy after 24 hours of immersion in a 0.1 M of $H_2SO_4$; Nyquist diagrams.

**Source: Progress in Organic Coatings (Elsevier).**

Palraj et al.[28] prepared an interpenetrating polymer network (IPN) of epoxy and PTh corrosion-resistant IPNs were prepared from immiscible resins (epoxy, silicone, and thiophene) using a cross-linking agent and a catalyst. The IPN formed of the two immiscible resins significantly enhanced the performance of coating material.

## 5.5   CP/NANOPARTICLE COMPOSITE-REINFORCED COATINGS

### 5.5.1 PANI/Nanoparticle-Reinforced Coatings

Nanocomposite coatings based on adding inorganic nanofillers in a polymer matrix are a new class of corrosion protection methods that show superior corrosion resistance and mechanical performance compared to conventional composite coatings.[29] Inorganic nanomaterials such as metal nano-powders, metal oxides, nano-glass flakes, nitrides and carbides, and nano-calcium carbonates have significant potential for enhancing the barrier performance of polymer coatings. A pure CP coating is often due to porosity, adhesion, and other problems, resulting in its single protective effect and poor durability. However, doping nonmetallic nanoparticles or metallic oxides are expected to improve these deficiencies on the basis of the original CP coatings.

PANI/nanocomposite coatings have shown excellent corrosion resistance. Extensive studies have been reported to improve the corrosion resistance of the PANI coatings by incorporating various nanoparticles, such as $TiO_2$,[30,31], $ZnO$[32], $SiO_2$,[33],$CeO_2$[34,35], and $Fe_3O_4$[36].

### 5.5.1.1 TiO₂ Nanoparticles

$TiO_2$ is the hottest material among the various nanometal oxides that was used to improve the performance of CP coatings, due to its unique charge carrier, oxidizing power, non-toxicity, and chemical and photo stability properties. A number of studies, in recent times, have focused on the application of $TiO_2$/conducting polymers nanocomposite coatings.[24,30,31,37,38]

$TiO_2$ can effectively reduce the permeability and improve the adhesion and corrosion capacity of the original PANI coating. The corrosion inhibition mechanism of $TiO_2$ nanoparticles/conducting polymers would be a complex one. The following factors would take charge of the enhancement of the polymer[24]: (1) $TiO_2$ nanoparticles could get trapped on coating pores and offer a physical barrier against external corrosive ions; (2) more doped anions would release due to more surface area induced by temper of $TiO_2$ during PANI polymerization; (3) p–n junctions would be formed in the interface of PANI/$TiO_2$ since PANIs are p-type and $TiO_2$ is an n-type semiconductors. Besides, currently, in recent times, photogenerated cathodic protection was also reported by combining the $TiO_2$/PANI nanocomposites coatings as photoresponsive coatings. Application of conducing polymer/$TiO_2$, as well as other semiconductor composites would be described in the following section.

### 5.5.1.2 ZnO Nanoparticles

ZnO nanoparticles are promising candidates to enforce the property of polymer. Compared to $TiO_2$ nanoparticles, ZnO nanoparticles have better optical and semiconducting (bandgap = 3.4 $eV$) properties and these properties are highly dependent on their size and shape.[19] Just as $TiO_2$, nano-ZnO/PANI can also construct a p–n junction at the interface of p-type PANI and n-type ZnO, and the formed junction would serve as a barrier against charge transport.

Numerous investigations[32,39–44] have been reported in the topics of ZnO highlighting the improvement in physical, electrochemical, and anti-corrosive properties of polymer nanocomposites. Reinforcement to an epoxy coating by PANI/ZnO has been confirmed by Mostafaei et al.[32], who successfully prepared conductive PANI/zinc oxide nanocomposites as the addition to the epoxy coatings. Compared with pure epoxy and epoxy/PANI coatings, the epoxy coatings contained ZnO/PANI particles have better barrier, adhesion, and corrosion resistance, with three orders of magnitude in corrosion rate. Meanwhile, PANI-ZnO in epoxy coatings significantly reduces the water permeability of the coating. Similar results were also noticed by Hu et al.[45], who prepared the poly(o-toluidine)/nano ZnO/epoxy composite coating on steel. The results showed that the composite coating containing a poly(o-toluidine)/nano ZnO composite has a higher corrosion resistance than that of poly(o-toluidine). The enhancement of corrosion protection ability of poly(o-toluidine)/nano ZnO composite containing a coating may be due to the formation of a more uniformly passive layer on a steel surface and the addition of nano ZnO particles, which increases the tortuosity of the diffusion pathway of corrosive substances.

### 5.5.1.3 SiO$_2$ Nanoparticles

Due to SiO$_2$ nanoparticles, these are the ideal candidate in the application of coatings, particularly epoxy coatings either as fillers or as inhibitor containers, due to the advantages of low cost, transparent, and scratch-resistance features. Additionally, mesoporous SiO$_2$-loading corrosion inhibitor storage is widely in the fabrication of self-healing coatings. The negatively charged OH$^-$, on the surface of SiO$_2$ colloidal particles, can facilitate in the synthesis of PANI through electrostatic interaction[46]. Therefore, a SiO$_2$ colloid can be used as a stabilizer to reduce the agglomeration of PANI. Recent reports indicated that the PANI/SiO$_2$ nanocomposite in reign could obviously enhance the corrosion resistance as well as adhesion of the reigns.[33,47–52]

Hydrophobic PANI-SiO$_2$ composite nanocomposites synthesized chemically and deposited on mild steel exhibit improved protective efficiency compared with PANI.[49] A corrosion test was performed in a 0.1 M sulfuric acid solution. The PANI/modified SiO$_2$ composite coatings showed excellent corrosion-protection properties. The synergistic effects of electrochemical activity of PANI and improved compact PANI-coating due to SiO$_2$ nanoparticles as well as the hydrophobicity of the hybrid PANI/modified SiO$_2$ coatings resulted in excellent corrosion-resistance performance.

### 5.5.1.4 CeO$_2$ Nanoparticles

Cerium dioxide (CeO$_2$) as a rare metal oxide has a variety of applications because of its special oxygen storage and release capacity, which is derived from two different valence states ($Ce^{4+}$, $Ce^{3+}$) with a unique reversible redox property, and plentiful oxygen vacancies.

Of all the rare metals, $CeO_2$ is also considered the most effective corrosion inhibitor and has attracted considerable interest to replace toxic chromates. Nanoparticles of $CeO_2$, like other metal nanoparticles, possess special physicochemical properties. They have a corrosion-inhibiting property, which is linked to possible complexation of species responsible for loss of passivation at a galvanized film surface.[24] Further, some smart coatings with a self-healing ability can be fabricated with $CeO_2$ nanoparticles.

Recently, Lei et al.[34] synthesized PANI/$CeO_2$ nanocomposite (NPs) by in-situ polymerization of aniline in the presence of $CeO_2$ nano-particles to improve the corrosion of epoxy coating (see Figure 5.3). TEM images in Figure 5.3 revealed sphere-like $CeO_2$ NPs in the range of 5–10 nm. A HRTEM image of the $CeO_2$ nanoparticles is given in Figure 5.3b, in which well-ordered stripes of nanostructures can be seen clearly, and the inter-planar spacing calculated to be 0.31 nm can be assigned to the (111) reflection plane of $CeO_2$. As shown in Figure 5.3c, apparently the $CeO_2$ nanoparticles were dispersed homogenously in the matrix of PANI. Figure 5.3d reveals the spongy nature of PANI with a

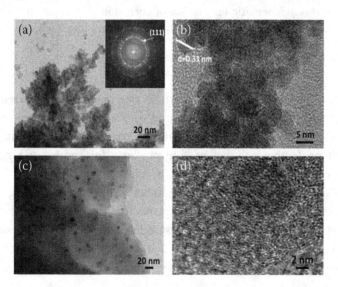

FIGURE 5.3  (a) TEM image of $CeO_2$ and (b) HRTEM image of $CeO_2$; (c) TEM image of PANI/$CeO_2$ and (d) HRTEM image of PANI/$CeO_2$.

**Source: Progress in Organic Coatings (Elsevier).**

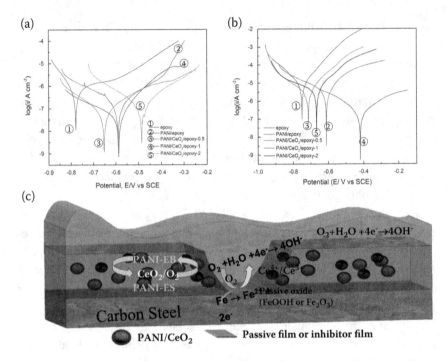

FIGURE 5.4 Potentiodynamic polarization curves for carbon steel coated with the control epoxy and the PANI/CeO₂/epoxy in a 3.5% NaCl for (A) 1 day and (B) 15 days. (C) Schematic illustration of corrosion protection mechanism for PANI/CeO₂ nanocomposites coatings on carbon steel in a 3.5% NaCl solution.

**Source: Progress in Organic Coatings (Elsevier).**

network-like structure, and the particle size obviously increased due to the encapsulation of PANI. Compared to the PANI/epoxy coatings, the coatings prepared from the PANI/CeO₂ nanoparticles exhibited more excellent protective and self-healing property (as shown in Figure 5.4). The exceptional improvement of corrosion protection performance of the PANI/CeO₂/epoxy coatings is associated with the synergetic protection of the enhancement of the protective barrier due to the role of CeO₂ nanoparticles and PANI against the diffusion of aggressive ions (e.g., Cl⁻) and the improvement of self-healing protection attributed to the redox behavior of PANI.

### 5.5.1.5 $Fe_2O_3$ Nanoparticles
Due to the environmentally friendly properties and low cost, $Fe_2O_3$ has also attracted wide attention in the field of catalysts, magnetic recording,

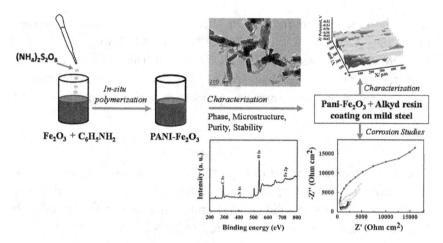

FIGURE 5.5   Schematic diagram of the research process.

**Source: Materials Chemistry and Physics (Elsevier).**

gas sensing, electrode material, and medical field. As an n-type semi-conductor, α-Fe2O3 combined with PANI was also reported to modify the property of reigns. Sumi et al.[52] applied a PANI-$Fe_2O_3$ composite in industrial alkyd resin and verified its protective effect on low-carbon steel; the research process is shown in Figure 5.5. The results show that good barrier performance and passivation effect are the basis of long-term protection of the composite coating on a metal substrate, and the complementary cathodic reaction observed in an acid medium provides a reference for the development of environmental protection and anti-corrosion coating.

### 5.5.1.6 Other Nanoparticles

Compositing is one of the modification techniques mentioned whereby conducting polymers are blended with themselves or with conventional polymers as well as with epoxy resins.

Beside the nanomaterials described above, the metal oxides such as $Al_2O_3$, CuO, and $ZrO_2$ were also applied with PANI in the anticorrosion field. Some materials with flake or layered structures have also been widely used in coatings. For example, clay, especially montmorillonite (MMT), is widely used in anti-corrosive coatings because of its high aspect ratio and excellent shielding properties. Besides, the materials of the graphene family, such as graphene, graphene oxide, and reduced graphene oxide, as well as other layered materials, e.g., $C_3N_4$, $MoS_2$, have attracted much attention in the field of anti-corrosion. Coatings, including PANI/Clay[53–55], PANI/GO[55–58], PANI/Graphene,[59–61] and PANI/MWNT[62], were also fabricated. The introduction of those nanomaterials into the interchain space of PANI can significantly improve the mechanical and anti-corrosion properties of the coating containing PANI.

## 5.5.2 PPy/Nanoparticle-Reinforced Coatings

Due to its excellent properties (e.g., stability in the oxidized form, good redox properties, great optical and electrical properties, and ability to provide high electrical conductivity) and simple synthesis method, PPy has been extensively studied compared to PANI.

Nanoparticles were also employed to strengthen the protective performance of the PPy and PTh. Nanoparticles including $SiO_2$[63], graphene(Gr)[60,64–66], graphene oxide (GO)[67,68], and MWNT[69], were reported previously. For example, Qiu and his colleagues[60] synthesized PPy-intercalated graphene on Q235 steel. They found that the corrosion performance of the composite coating containing a 0.5 wt% Gr showed a superior anti-corrosive ability. Also, they suggested that the possible mechanism for synergistic protection was based on a self-healing PPy and impermeable graphene sheets that were confirmed by the observation of corrosion products beneath the coatings.

Zhu et al.[70] reported a promising novel composite coating based on a compound pigment of zinc phosphate and PPy functionalized graphene (ZGP), two ratios of PPy functionalized graphene oxide (GO-PPy) nanocomposites were prepared by in-situ polymerization of pyrol (Py) on the surface of GO. Embedding a small amount of a ZGP-2 composite pigment into the waterborne epoxy coating significantly improved corrosion protection of the prepared composite coating. As can be seen in Figure 5.6, the schematic represents the corrosion protection of ZP and ZGP-2 composite coatings. Because of the π-πbonds between GO and PPy, the functionalized GO nanomaterials had no direct connections and dispersed homogeneously in a waterborne epoxy resin matrix that produced an excellent barrier effect.

Fekri et al.[71] reported that ZnO nanoparticles increase the PPy electrical conductivity. Due to an increase in the electrical conductivity of the PPy coatings by loading ZnO nanoparticles, obvious enhancement in the corrosion protection performance of the PPy coatings on an Al alloy 2024 were observed by loading ZnO nanoparticles.

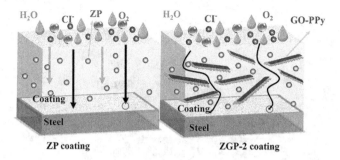

FIGURE 5.6   Schematic representation of the corrosion protection of ZP and ZGP-2 composite coatings.

**Source: Composites Part A: Applied Science and Manufacturing (Elsevier).**

Jadhav et al.[72] studied the anti-corrosion performance of a PPy/ $Fe_2O_3$ nanocomposite coating. They found that a PPy/$Fe_2O_3$ coating showed a better corrosion resistance than a $Fe_2O_3$ coating. It was shown that the ability of a PPy for passivation and interaction of a PPy with ions was responsible for its higher corrosion protection.

## 5.6 ANTI-CORROSION MECHANISM OF CP COATINGS

As an anti-corrosion material, CP coatings have the unique corrosion protection mechanisms. Nobling or anodic passivation, barrier protection, mediation of oxygen reduction, cathodic protection, and controlled inhibitor release mechanism are some of the protection mechanisms proposed for CP-coated metals and alloys. The following anti-corrosion mechanisms were proposed to understand the different corrosion-protective phenomenon of CPs.[73]

### 5.6.1 Anode Protection Mechanism

This mechanism is the most widely known and most important anti-corrosion mechanism of conductive polymer anti-corrosion materials. It was first proposed by Deberry in 1985.[9] Numerous studies have shown that the conductive polymer coatings can oxidize the protected metal substrate through the redox reaction during the service process, forming a dense oxide layer at the interface to passivate the metal and prevent corrosion. The further contact between the dielectric medium and the metal substrate can effectively prevent the occurrence of corrosion.[73,74] A schematic illustration of this mechanism is presented in Figure 5.7. More detail about the anodic mechanism can be seen in Chapter 2.

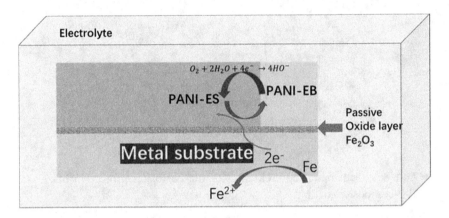

FIGURE 5.7 The concept of an ennobling (anodic protection) mechanism using CPs.

**Source: created by author for this publication.**

### 5.6.2 Physical Barrier Mechanism

The conductive polymer anti-corrosion coating possesses compact components and good adhesion to the metal, which can effectively isolate a corrosive medium (water, oxygen, corrosive ions, etc.). At this time, in the primary cell system formed in the entire etching environment, the reduction reaction that occurs in the cathode will be transferred from the metal/conductive polymer interface to the conductive polymer/corrosion environment interface; the transfer of the reduction reaction site can reduce the reduction product at the metal/conductive polymer interface, [75,76] further inhibiting the coating foaming and delamination or peeling from the metal surface. It is also important to maintain the excellent physical barrier properties of the conductive polymer to improve the corrosion resistance of the conductive polymer.

### 5.6.3 Dopant Release Mechanism

The doped conductive polymer is in the oxidation state with conductive properties. In its working environment, the reduction of the conductive polymer and the ion exchange between the coating and the corrosion environment will be accompanied by the de-doping of dopant anions. Some special doping ions, such as phosphomolybdate and sulfonic acid, can be complexed with the metal cations on the protected metal surface, inhibiting the further corrosion of the metal. [76–78] According to this, the anti-corrosion efficiency of the coating can be improved by introducing a

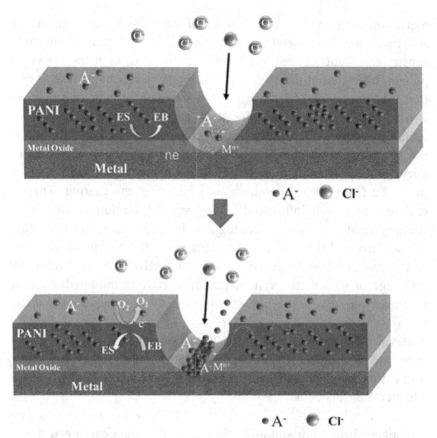

FIGURE 5.8 Controlled inhibitor release mechanism for a metal, M, coated by a CP layer such as PANI doped with an ion, A-, which acts as a corrosion inhibitor.

**Source: created by author for this publication.**

corrosion inhibitor into the conductive polymer anti-corrosive coating, and the self-repair of the micro-defects in the coating and metal substrate can be realized by releasing the corrosion inhibitor during the service of the coating. When a scratch exists on the CP-coated metal, there is a galvanic coupling between the metal and the CP (Figure 5.8). The cathodic reaction involves the reduction of the CP associated with the release of the doping anions, whereas the anodic reaction involves the oxidation of the metal.[73]

## 5.6.4 Electric Field Shielding Mechanism
It is reported that a doped-state CP coating having a semiconductor or a conductive property may form an electric field between the protected

metal substrate to limit the transfer of electrons between the metal and its oxide form, thereby inhibiting further corrosion of the metal substrate,[73] but conventional organic coatings such as epoxy or polyurethane cannot form such an electric field. It is important to note that the process of dedoping the ions when the conductive polymer anti-corrosion coating is reduced is also accompanied by an ion exchange process.[79] Anions such as chloride, sulfate, fluoride, etc. in the corrosive environment may be doped into the polymer chain during this process, and some will also migrate to the surface to be protected from pitting, resulting in a failure of the coating. The ion exchange process is influenced by the types of the doping ions in the coating, the thickness of the coating, the hydrophobicity of the coating, the porosity and the like, and especially when the dopant which is not easily dedoped is introduced in the conductive polymer chain, the exchange of ions in the system can be effectively inhibited, so as to improve the corrosion-resistant effect of the coating on the metal substrate.

Based on doping/dedoping of CP coatings, the ionic selective anti-corrosion mechanism of a conductive polymer was also proposed. Effects of the doped ions on the corrosion performance of CPs can be found in Chapter 2 of this book.

For example, Pour-Ali et al.[78] doped PANI with a macromolecular organic acid camphor sulfonic acid (CSA), and then combined it with epoxy (Epoxy) to prepare a composite coating for carbon steel corrosion protection in a NaCl environment. The doping of CSA would effectively reduce the ion exchange between the doped anions and Cl$^-$ in a solution, and thus restrain the diffusion of Cl$^-$ to the coating. Further, compounding with anepoxy resin can significantly reduce the defects of PANI itself and improve the physical barrier effect of the coating, so it can effectively block the inward diffusion of Cl$^-$. As shown in Figure 5.9, PANI can form a passivation layer at the coating/carbon steel interface. Partially dedoped CSA can form insoluble compounds with Fe$^{2+}$ ions released from the surface of carbon steel during a long-term service. These insoluble compounds can fill the defects of the passivation layer and the coating, and strengthen the coating's barrier to corrosive ions. Therefore, a PANI-CSA-Epoxy composite coating can provide long-term and stable corrosion protection for carbon steel.

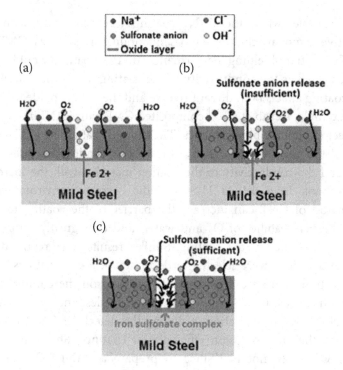

FIGURE 5.9  Schematic of the anti-corrosion mechanism of PANI-CSA-Epoxy coatings. (a) epoxy coating without PANI, (b) epoxy coating with low amount of PANI, and (c) epoxy coating containing sufficient PANI.

**Source: Corrosion Science (Elsevier).**

## 5.6.5 Hybrid Synergistic Mechanism

Synergistic effect of organic/inorganic hybrids should be emphasized here.

Herein, we would like take a PANI coating as an example to analyze the synergistic effect on corrosion protection of PANI coatings. As previous described, the PANI coatings have many advantages. One fatal defect is that it is a porous structure, and the corrosive ions can enter the coating/metal interface via these micro-pores to some extent weaken the anti-corrosion effect of the coating on the metal substrate. In order to solve this problem, the research of PANI coating at this stage is combined with other materials to improve the anti-corrosion performance of the coating.

Interestingly, we need to note that the mechanisms involved in CPs are not independent, but related with each other. Thus, a hybrid synergistic mechanism was proposed to understand the effect of inorganic nanoparticles on coating corrosion behavior.

For example, when the PANI coatings are reinforced with some conductive nanoparticles or nano-semiconductor, such as CNTs and ZnO, $CeO_2$, the shielding performance of the coating would be improved, as well as the conductivity of the coating; while the conductivity of the coating affects both dopant release and the anodic oxide behaviors.

For example, Deyab et al.[80] composited CNT with PANI into coatings to protect aluminum bipolar plates. They found that as the CNT content increases, the porosity and defects of the PANI coating decreases significantly, and the conductivity of the coating increases with the increase of CNT content. In a 0.1 M $H_2SO_4$ acidic corrosive environment, the introduction of CNT can increase the barrier of the coating to $SO_4^{2-}$, reduce the permeability of $O_2$ and water, and thus greatly improve the corrosion resistance of the coating. Similar results were reported when $CeO_2$ nanoparticles were selected to decorate the PANI coatings.[34]

In addition, there are many reports focusing on the combination of PANI with metals, metal oxides, non-metal oxides, and other materials for metal corrosion protection, reducing the defects of PANI itself, and improving the anti-corrosion efficiency of coatings. Shi et al.[33] combined PANI with modified silica to prepare a PANI-$SiO_2$ composite coating with hydrophobic properties. Studies have shown that the modified $SiO_2$ can effectively fill the pores of PANI to form a dense coating structure. At the same time, the coating has a certain self-repairing function. After soaking in a 0.1 M $H_2SO_4$ environment for 71 days, the corrosion resistance of Q235 carbon steel is still as high as 99.99%. Pagotto et al.[31] combined PANI with titanium dioxide nanotubes to prepare PANI/n-$TiO_2$ coatings for corrosion protection of carbon steel and welded carbon steel. However, when the thickness of the coating reaches 4.5 μm, the porosity of the coating will increase due to the increase of mechanical tension, which will reduce the anti-corrosion efficiency.

## 5.7 SEVERAL APPLICATIONS OF PANI COATINGS

Among all kinds of conductive polymers, a PANI coating is the most widely used, so this section only describes several applications of PANI coatings.

### 5.7.1 Anti-Corrosion for Oil Storage Tanks and Pipelines

The materials of the various parts of the oil tank are different. When they come into contact, a potential difference is formed on the surface of the oil tank. Moreover, the oil stored in the tank contains some electrolytes, and these electrolytes will electrochemically corrode the storage tank under the effect of the surface potential difference. Adding some conductive filler to the coating on the inner wall of the tank can eliminate the potential difference on the tank. People often use inorganic small molecule particles as conductive fillers. However, the structures of organic high polymers and inorganic small molecule particles are not consistent. This makes the dispersibility between the inorganic filler and the polymer matrix poor, and it is difficult to form a stable homogeneous system. Ultimately, the conductivity of the coating is weakened and the corrosion resistance of the coating is reduced. As a conductive polymer material, PANI has a structure similar to that of the organic components in the coating, good compatibility, and strong coating stability. At the same time, PANI has a certain electrical conductivity and can replace inorganic fillers for anti-corrosive coatings on oil tanks.[81]

The use of coating to isolate the corrosion medium from the pipe can effectively prevent the corrosion of the pipeline. The antic-orrosion in sewage treatment pipeline and oil pipeline is the main application of a PANI coating in pipeline anti-corrosion. CORRPASIVE4900, developed by Ormecon in Germany, is mainly used in municipal sewage treatment systems. Its primer contains dispersed PANI and its topcoat is epoxy resin.[82]

## 5.7.2 Zinc-Rich Epoxy Primer

A marine coating is a form of protective coating mostly used in the marine atmosphere to protect ships, yachts, vessels, tankers, and other materials from the salty water of the sea. Because of its specific functional properties, a marine coating can deliver a greater level of protection to the surfaces to which it is applied.

A zinc-rich epoxy primer (ZRP) has been extensively utilized to protect steel substrates in many industrial and marine environments since the 1930s.[83] In these coatings, the polymer matrix establishes a barrier against aggressive species, and the addition of a high amount of zinc particles to the epoxy resin provides a coating system with cathodic protection. The zinc content of epoxy zinc rich primer is usually higher than 65 wt%.

Numerous studies on the mechanism of ZRP have indicated that the zinc particles' sacrificial action occurs when there is continuous contact between the metal particles.[84,85] ZRPs exhibit good sacrificial protection during the early stages of their service life. However, because of the sacrificial reaction of zinc, zinc oxide or hydroxides would be formed outside, resulting in a considerable increase in the ohmic resistance between electroactive sacrificial zinc particles and the underlying metallic substrate.[84,86] Therefore, the sacrificial cathodic protection of the coating is gradually weakening. Also, corrosion products of zinc can fill the coating pores and thus reduce the duration of its cathodic protection. Consequently, the improvement of the effective service life of the zinc-rich protective coatings is an important issue from diverse viewpoints, such as the views of industry, economic cost, and the environment.[87]

There are two main ways to alleviate this problem. One is to add conductive fillers, such as carbon black, aluminum powder, conductive polymers, carbon nanotubes, and so on, to improve the internal conductivity of the coating. It is understood that the defects and porosity of the coating would facilitate the percolation of the aggressive anions into the coating, which results in the corrosion of zinc particles. Another way is to add shielding fillers, mainly lamellar fillers, that can form labyrinth effects inside the coating, extend the diffusion path of corrosive media, and slow down the consumption rate of zinc. From the view of the above statement, attempts have been made to enhance the cathodic protection duration of zinc-rich coating through inclusion of different types of fillers and additives. In this regard, abilities of zinc nanoparticles,[87] aluminum pigments,[88] carbon black,[84] and carbon nanotubes[89] to decrease the initial contact electronic resistance and boost sacrificed zinc utilization have been explored. With the rise of new materials such as CNTs and Gr, the excellent electrochemical performance and corrosion resistance of these materials are gradually being discovered. Recently, Hayatdavoudi et al.[90] studied the enhanced cathodic protection properties of a zinc-rich epoxy for corrosion protection of carbon steel by addition of a small amount of graphene.

Besides the above-mentioned conductive fillers, CPs, e.g., PPy or PANI, are alternative additive conductive fillers, have also attracted considerable attention. For example, Meroufel et al.[91] employed hydrochloride PANI powder as the conductive filler into ZRP. However, analysis of this innovative powder formulation indicated no improvement in coating conductivity, but an excellent barrier property was obtained. Akbarinezhad et al.[54] applied PANI and PANI-clay in ZRPs to extend the long-term corrosion protection of ZRPs.

Liu et al.[86] investigated the mechanism of PANI-grafted graphene (PANI-Gr) nanosheets on the corrosion performance of ZRP coatings.

The PANI/Gr nanosheet (Figure 5.10b), synthesized from in-situ chemical oxidation, method was introduced into an epoxy zinc-rich coating (Figure 5.10a). As shown in Figure 5.10c, the conductivity of the PANI nanosheets was much improved due to the addition of Gr. An accelerated corrosion test was performed by introducing artificial defects into the coating. The potential changes of the artificial defect-included samples and localized EIS of the defect area were conducted, to investigate the impact of the PANI-Gr fragments on corrosion process. The results show that graphene can play a bridging role in epoxy zinc-rich coatings, connecting zinc particles in series to improve the cathodic protection effect of the coatings. And the electrochemical results indicated that a mixture of Gr at a content of 0.6% enhanced the sacrificial cathodic protection of ZRP coating due to extending its active sacrificial period.

Ramezanzadeh et al.[59] combined the composite nanoparticles of GO and PANI with zinc-rich epoxy resin (ZRC) to prepare a composite coating with cathodic protection for the corrosion protection of steel. In a 3.5 wt.% corrosive environment, galvanic couples are formed between zinc and the steel substrate. Zinc as an anode can suppress the corrosion of the base steel plate as a cathode. The addition of GO reduces the porosity defects of the coating and improves the coating resistance. At the same time, the addition of PANI promotes the electrical contact between zinc and reduces the oxidation of zinc, thereby extending the cathodic protection effect of the coating on the protected substrate.

## 5.7.3 Marine Anti-Biofouling Coatings

Iron and steel are the main materials used in offshore facilities and equipment. However, steel can be corroded in complex marine environments. Marine microorganisms can cause local corrosion of offshore facilities and equipment. Some heavy metal ions are often added to traditional heavy anti-corrosive coatings. This is because these heavy metal ions can solidify microbial proteins and kill microbes adsorbed on the coating. However, the heavy metals in the coating will precipitate into the ocean, which will seriously pollute the environment. The special properties of PANI make it possible to reduce the release of pH and inhibit the adsorption of microorganisms on the surface of iron and steel, which are suitable for alkaline environment, so as to achieve the dual objectives of anti-fouling and anti-corrosion.[92,93]

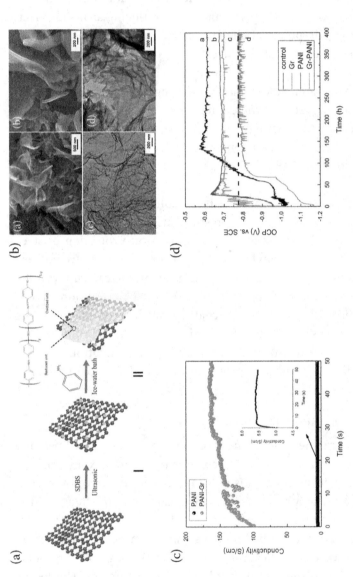

FIGURE 5.10 (a) The synthesis procedures of PANI-Gr. ( I ) Gre were first modified with SDBS, (II) PANI were synthesized on the surface of the modified Gr; (b) SEM image of the synthesized PANI-Gr; (c) Direct conductivity of the PANI and PANI-Gr; (d) OCP of the DH 32 steel coated with ZRP (control) (line a), Gr-ZRP (line b), PANI-ZRP (line c), and PANI/Gr-ZRP (line d) with artificial defects in a 3.5% NaCl solution.

**Source: created by author for this publication.**

Recently, Cai et al. reported a one-pot method to synthesize dedoped bromo-substituted PANI (Br-PANI) for anti-fouling and anti-corrosion applications.[80] Three Br-PANI polymers were prepared with Br contents of 22.10, 42.51, and 46.29 wt%. After immersion in an NaCl solution (simulate the marine environment) for 100 days, the Br-PAIN/epoxy coatings showed better anti-corrosion capabilities than pure epoxy, and the anti-corrosion properties improved with increasing Br content. This BrPANI has the potential to be incorporated into different coatings to enhance their anti-fouling and anti-corrosion properties.

Details about the PANI anti-fouling coating applications can be referred to in the following chapter.

### 5.7.4 Waterborne Anti-Corrosive Coatings

With the deterioration of global environmental pollution and the intensification of greenhouse effect, governments, organizations, enterprises, and researchers from different countries have paid more and more attention to waterborne coatings. Reduction of VOCs such as organic solvents is a key challenge in the paint industry due to the increased environmental and health legislations.[81] VOCs participate in atmospheric photochemical reactions to create health hazards. Development and use of waterborne coatings have reduced VOC in traditional alkyd paints.

Water-based coatings emit less VOCs than its substitutes, such as solvent-based coatings. The global waterborne coatings market is driven by the rise in demand for architectural coatings. According to the waterborne coatings market, analysis published by transparency Market Research in May 2022, the global waterborne coatings market was valued at US\$77.4 billions in 2018 and is anticipated to expand at a compound annual growth rate (CAGR) of 4.9% from 2019 to 2027.[94]

Waterborne PANI dispersion has had extensive attention due to its environmental friendliness and good processability, whereas the storage stability and mechanical property have been the challenge for the waterborne PANI composites.

The application of PANI in water-based anti-corrosion coatings is not only to combine it with water-based coatings, but also to maintain its corrosion resistance on the basis of improving the compatibility of PANI with water solvents and water-based coatings. As far as PANI processing is concerned, an organic solvent-based system dominates in which conducting PANIs are dispersed or dissolved in toxic solvents like m-cresol and chloroform, or volatile organic solvents like xylene or toluene, which causes serious environmental pollution and potential human health damage.[95,96]

The most important is to synthesize waterborne PANI to improve the solubility and dispersibility of PANI in water. Various approaches have been reported to prepare waterborne PANIs, including self-doping, block and graft modification of PANI, preparation of watersoluble PANI by aqueous emulsion polymerization, template polymerization method, and dopant induction solubilization.

The incorporation of appropriate substitutes, such as sulfonic acid, boronic acid, and carboxylic groups, at the phenyl rings or nitrogen sites of PANI have turned out to be a simple and effective methodology to improve the solubility of PANI.[97] In addition, waterborne PANI dispersion with good stability could be prepared through in-situ synthesis of PANI using cellulose nanocrystals[98] or GO.[99]

For example, Zhu et al.[99] employed GO and PANI to modify the waterborne epoxy coatings (WEP). PNAI was first in-situ synthesized from the surface of GO to form the PANI/GO composites, which showed the excellent water dispersion ability in water. Then the composites was blended into the waterborne epoxy reign. Compared with pure waterborne epoxy (WEP) coating, the prepared GO/PANI/WEP coatings exhibited remarkably reinforced anti-corrosion performances, and the superior anti-corrosion performances of GO/PANI/WEP coatings were mainly attributed to the synergistic effect of GO as the physical barrier and PANI as the CPs, as illustrated in Figure 5.11.

Morsi et al.[100] synthesized self-doped PANI by using aniline and 4-amino benzenesulfonic acid as monomers. Moreover, surfactants are also utilized for the stabilization of waterborne PANI and PPy dispersion, such as poly (N-vinylpyrrolidone), methylcellulose, and polyacrylic acid. Further, the self-doping approach has been widely used to prepare water-based PANIs, in which the sulfonic or phosphoric acid group linking onto the aromatic ring or nitrogen atom[101] served as hydrophilic segment to provide water soluble or dispersible properties. Chan et al.[102] dissolved the eigenstate PANI in dimethyl sulfoxide, reduced it with NaH, and reacted with propylsulfonic acid to obtain N-substituted waterborne PANI. Zhang et al.[103] prepared self-doped waterborne conducting PANIs through the co-polymerization of 2-sulfoaniline and aniline using potassium hexacyanoferrate (III) as an oxidant, in which the doped state is directly obtained with hexacyanoferrate (II) anion and sulfonate group surrounding imine and amine atoms in a PANI backbone.

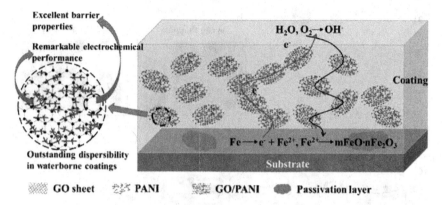

FIGURE 5.11   A schematic diagram to illustrate the GO/PANI synergetic effects in anti-corrosion mechanism of the WO/PANI/WEP.

**Source: Progress in Organic Coatings (Elsevier).**

The co-polymerization of aniline derivatives and aniline is also an alternative approach to prepare self-doped and water-soluble PANI derivative. Co-polymerization of aniline and aniline derivatives with hydrophilic groups is another important way to obtain waterborne conductive PANI. Hua and Ruckenstein[104] co-polymerized aniline with low molecular weight polyethylene oxide (PEO)-grafted 3-aminophenol (AP-g-PEO) to obtain a new type of water-soluble copolymer.

The method of aqueous emulsion polymerization has also been used to improve the water solubility of PANI. The specific process is to add a water-soluble steric stabilizer to the polymerization system of aniline, so that the polymerization of aniline occurs inside the micelle to form a colloidal dispersion of PANI. Rao et al.[105] used ammonium persulfate and benzoyl peroxide as oxidants, and changed the ratio of aniline to polyvinyl alcohol (PVA), respectively, obtained water-soluble conductive polymers with mass fractions of PVA of 10, 30, 50, and 70 by emulsion polymerization and inverse emulsion polymerization.

Wang et al.[97] prepared PANI-g-EPVA nanocomposites through chemical graft polymerization, and the preparation scheme is shown in Figure 5.12a. In order to make a comparison between a conventional preparation method and a chemical graft polymerization, an in-situ PVA/PANI nanocomposite is also prepared through the oxidation polymerization of aniline in a PVA solution. The optical photos of in-situ PVA/PANI and PANI-g-EPVA dispersions and the corresponding films are presented in Figure 5.12b. It is found that a PANI-g-EPVA film displays

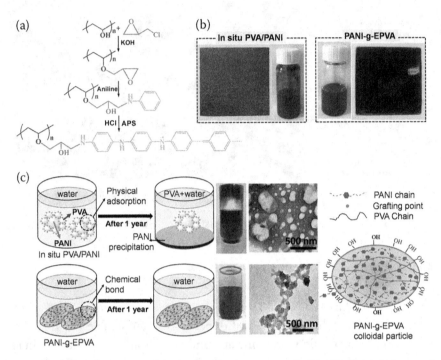

FIGURE 5.12   (a) Preparation scheme of PANI-g-EPVA. (b) The optical photos of in-situ PVA/PANI and PANI-g-EPVA dispersions and the corresponding films. (c) The molecular model and stabilizing mechanism of in-situ PVA/PANI and PANI-g-EPVA dispersions, as well as the optical photos and TEM morphology.

**Source: Scientific Reports (Elsevier).**

more uniform and smooth morphology compared with an in-situ PVA/PANI film. And macroscopic PANI precipitation was initially generated inside in-situ PVA/PANI dispersion after storage for 15 days, and the amount of PANI precipitation increases with prolonging the storage time. In contrast, the PANI-g-EPVA dispersion keeps stable for more than one year, as presented in Figure 5.12c. Particle agglomerations are detected in the TEM morphology of in-situ PVA/PANI dispersion, and the particle size distributes unevenly, while the particle size of PANI-g-EPVA dispersion is more uniform.

Waterborne conducting PANI can also be prepared using template polymerization. The unsaturated monomer is first compounded with the template polymer, and then polymerized on the template, and the finally formed polymer and the template form a complex. Li et al.[106] synthesized water-soluble PAA/PANI nanocomposites by a template-directed method using high molecular weight polyacrylic acid PAA (250000 MW) as a template.

In the strategy of dopant-induced solubilization, water is often used as a solvent or dispersion medium, and a protonic acid with a hydrophilic functional group was chosen as a dopant to prepare an aqueous solution or aqueous dispersion of doped PANI. Wang et al.[107] used acid phosphate or sulfonate with long-chain hydrophilic groups as counter-ion dopants to doped PANI to obtain water-soluble or water-dispersible conductive PANI. Phosphate end-group doped PANI polyethylene glycol chains are water soluble. Experiments show that when the polyethylene glycol chain is long enough (Mn=550), the doped PANI is completely soluble in water.

Subsequently, numerous PANI waterborne coatings were fabricated through hydrophilic anions doping. For example, Gurunathan[95] reported a castor oil-based waterborne PU/PANI (named COWPU/PANI) showed that the thermal stability of PU was improved by the presence of PANI. COWPU/PANI hybrid dispersions were synthesized with PANI-DBSA by a self-doping method to prepare different conductive composites. The CP blend film can be used as antistatic and corrosion inhibitor materials. Zhu et al.[108] prepared water-dispersed self-doped sulfonated PANI (SPANI) and self-doped carboxylated PANI(CPANI), which were prepared by the co-polymerization of aniline with m-aminobenzenesulfonic acid and m-aminobenzoic acid, respectively. The resultant SPANI and CPANI nanofibers were used as anti-corrosion pigments for water epoxy coatings (EP), which have better corrosion resistance compared to waterborne epoxy coatings.

Qiu et al.[109] synthesized self-doped PANi (SPANi) nanofibers with an average diameter of 45 nm and length of 750 nm by co-polymerization of aniline and 2-aminobenzenesulfonic acid. The resulting nanofibers had excellent aqueous solubility, good conductivity up to 0.11 S/cm and showed reversible redox activity, making them suitable as corrosion inhibitors for waterborne anti-corrosive coatings. Owning to the ultra-small nanofiber structure and unique reversible redox behavior of SPANi, SPANi nanofiber incorporation at 0.5 wt.% would significantly improve the anti-corrosion capacity of the waterborne epoxy coating, results of which was supported by the potentiodynamic polarization and EIS results in Figure 5.13.

Beside the self-doping method to fabricate PNIA-based waterborne paint, PANI coupled with other nanoparticles were also intensively

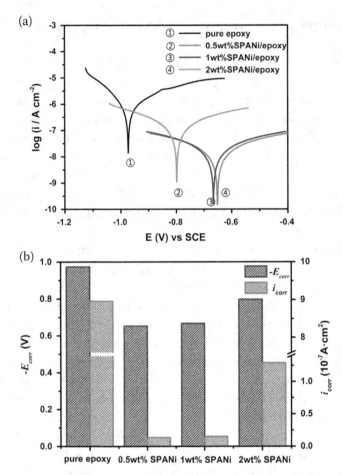

FIGURE 5.13 (a) Polarization curves and (b) corrosion parameters of pure epoxy; 0.5 wt% SPANI/epoxy; 1 wt.% SPANI/epoxy; 2 wt.% SPANI/epoxy coatings on the Q235 steel electrodes immersed in a 3.5% NaCl aqueous solution for 36 days.

**Source: Applied Surface Science (Elsevier).**

reported. Cai et al.[110] prepared the anti-corrosion coating based on a PANI/Gr composite filler and waterborne PU. The results showed that the new anti-corrosive coating increased the tortuosity of the electrolyte penetration path, and showed good corrosion resistance in the salt spray test at 120 hours. Later, using the same in-situ polymerization method, Xiao et al.[111] successfully prepared a more environmentally friendly waterborne zinc-based coating by adding a PANI graphene oxide composite. The composite materials prepared by in-situ polymerization can help PANI better embed in graphene oxide layer or adsorb on the surface of graphene oxide, which further promotes the uniform dispersion of PANI. Compared with the control group that only added PANI, the former has greater improvement in dispersion, conductivity, cathodic protection, and barrier performance, which makes PANI graphene oxide composite more suitable as a corrosion inhibitor of zinc-based water-borne anti-corrosion coating.

### 5.7.5 Proton Membrane Fuel Cell Bipolar Plate

Hydrogen is used as fuel of proton exchange membrane fuel cell (PEMFC). The only emission is pure water, which is one of the most promising new energy sources for automobile, train, and other means of transportation. At present, the high cost is the main problem that restricts the mass production and commercial application of PEMFC, especially the graphite bipolar plate. A graphite bipolar plate is applied for gas passage and expenditure accounts for about 60% of the entire battery pack. The extreme brittleness and low strength of the bipolar plate limits the increase of battery size and volume power density. Therefore, the search for alternative materials for graphite bipolar plates is becoming an urgent key problem to be solved in the process of PEMFC industrialization. The ideal bipolar plate material should not only have the characteristics of high conductivity, high mechanical strength, strong corrosion resistance, and good gas permeability resistance, but also meet the requirements of low cost, easy processing, and good dimensional stability. In addition, the bipolar plate should also be a good conductor of heat to ensure the uniform distribution of temperature in the battery stack and the smooth discharge of waste heat.[112]

316 stainless steel is used as a PEMFC bipolar plate to highlight the cost and performance advantages, but it still faces two technical problems in order to realize its application: first, the spontaneous formation of $Cr_2O_3$ passivation film on stainless steel surface will increase the contact resistance between the bipolar plate and gas diffusion layer; second, metal ions such as corrosion dissolved Cr, Fe, and Ni of stainless steel will reduce the activity of electrocatalyst and the conductivity of Nafion film. These problems will seriously affect the efficiency of PEMFC. Most of the researchers believe that the untreated stainless steel plate is not suitable for PEMFC directly while the use of PANI coatings on the bipolar plate could effectively solve this problem.

Huang et al.[113] electrodeposited a PANI coating on the surface of a 1Cr18Ni9Ti stainless steel bipolar plate by pulse voltage method. The working environment of PEMFC at 80°C was simulated, and $H_2$ was added into the system to study the corrosion resistance of a PANI-protected bipolar plate in an acid corrosive environment. It is found that the self-corrosion potential of stainless steel bipolar plate increases from 350 mV to 250 mV under the action of a PANI coating. A series of electrochemical performance tests show that the coating performance is stable and no degradation occurs during the test.

In order to reduce the porosity of the electrodeposited PANI coating on the 316L stainless steel surface and enhance the conductive properties of the coating, Zhang et al.[114] deposited ultra-low-load gold nano-particles on the PANI coating to obtain an AuNP-PANI composite coating. As shown in Figure 5.14a, gold nanoparticles can fill the pores of

(a)
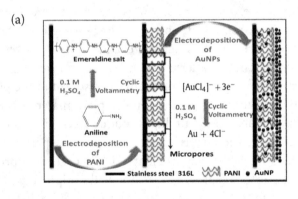

(b)
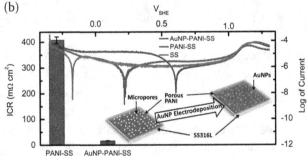

FIGURE 5.14 (a) Schematic showing the coating process of AuNP-PANI hybrid coating on SS316L; (b) potentiodynamic polarization curves of the three samples in a 1 mM $H_2SO_4$ solution at 80°C with inset showing the AuNP-PANI-coated SS316L.

**Source:** Acs Sustainable C (American Chemical Society).

PANI, effectively enhancing the barrier effect of the coating on corrosive ions, and making the coating hydrophobic. At the same time, the introduction of gold nanoparticles can greatly improve the conductivity and heat resistance of the coating, so the contact resistance of the composite coating system is significantly reduced compared to the PANI coating alone (Figure 5.14b). Potentiostatic polarization and potentiostatic polarization tests have shown that the AuNP-PANI composite coating system has excellent corrosion resistance, and has a good application prospect on 316L stainless steel bipolar plates of PEMFC.

## 5.8 SUMMARY AND OUTLOOKS

CPs and their family of stimuli-actuated polymers have proven to be invaluable in the field of anti-corrosion, and PANI and its derivatives have received wide attention as representatives of conductive polymers in the past two decades. Conductive polymers represented by PANI and PPy can be used in anti-corrosion coatings by ways of direct electrochemical deposition or chemically polymerized and blended in various resins. The organic-inorganic composites can not only improve coating anti-corrosive properties, but also improve the stability of the CP-contained coatings.

Several mechanisms have been proposed to understand the protection, such as anodic passivation, barrier protection, mediation of oxygen reduction, electric field shielding mechanism, and dopant release mechanism. Meanwhile, a hybrid synergistic mechanism was proposed to understand the improvement of a CP-based coating by nanoparticles.

Due to the excellent protective properties of PANI-contained coatings, the PANI coatings have a wide range of applications in oil and gas storage tanks or pipelines, marine fields, the waterborne coatings as well as proton membrane fuel cell bipolar plates.

Especially, waterborne PANI dispersion has had extensive attention due to its environmental friendliness and good processability, whereas the storage stability and mechanical property have been the challenges for the waterborne PANI composites. Various approaches have been reported to prepare waterborne PANIs, including self-doping, block and graft modification of PANI, preparation of water-soluble PANI by aqueous emulsion polymerization, template polymerization method, and dopant induction solubilization.

In general, although conductive PANI coating research has made great progress recently, there are still some challenges to overcome. For example, it is still a big challenge to find the best preparation method for a PANI- or PPy-containing composite with a long-term stability in a specific corrosion environment. In the future, it will be meaningful to explore or to fabricate multi-functional coatings, which can proceed multiple properties, such as anti-corrosion, anti-fouling, anti-icing, photothermal, and photocatalytic properties. Thus, further research is still required to fully display their charming potential.

## REFERENCES

1. Phull, B. 2010. In *Shreir's Corrosion*, ed. Cottis, B. *et al.*, 1107–1148. Elsevier Press.
2. Li, Y. & Ning, C. 2019. Latest research progress of marine micro-biological corrosion and bio-fouling, and new approaches of marine anti-corrosion and anti-fouling. *Bioactive Materials*, **4**, 189–195.
3. Hou, B. R. & Lu, D. Z. 2018. Corrosion cost andp reventive strategies in China. *Bulletin of Chinese Academy of Sciences*, **33**, 601–609.
4. Wang, X. H., Li, J. & Zhang, J. Y. *et al.* 1999. Polyaniline as marine antifouling and corrosion-prevention agent. *Synthetic Metals*, **102**, 1377–1380.
5. Zou, C. Y., Zhou, Q. H., Wang, X. H., Zhang, H. M. & Wang, F. S. 2021. Cationic polyurethane from $CO_2$-polyol as an effective barrier binder for polyaniline-based metal anti-corrosion materials. *Polymer Chemistry*, **12**, 1950–1956.
6. Liu, J. R., Lei, Y. H., & Qiu, Z. C. *et al.* 2018. Insight into the impact of conducting polyaniline/graphene nanosheets on corrosion mechanism of zinc-rich epoxy primers on low alloy DH32 steel in artificial sea water. *Journal of The Electrochemical Society*, **165**, C878–C889.
7. Hao, Y., Zhao, Y., Yang, X., Hu, B., Ye, S., Song, L. & Li, R. 2019. Self-healing epoxy coating loaded with phytic acid doped polyaniline na-nofibers impregnated with benzotriazole for Q235 carbon steel. *Corrosion Science*, **151**, 175–189.
8. Chakraborty, R., Manna, J. S. & Saha, P. 2019. Development and relative comparison of polypyrrole-calcium phosphate composite coatings with differential concentration of chlorophyll functionalized polymer particle achieved through pulsed electro deposition. *Surface and Coatings Technology*, **363**, 221–235.
9. DeBerry, D. W. 1985. Modification of the electrochemical and corrosion behavior of stainless steels with an electroactive coating. *Journal of the Electrochemical society*, **132**, 1022.
10. Mrad, M., Amor, Y. B., Dhouibi, L. & Montemor, F. 2017. Electrochemical study of polyaniline coating electropolymerized onto AA2024-T3 aluminium alloy: Physical properties and anticorrosion performance. *Synthetic Metals*, **234**, 145–153.

11. Wang, T. & Tan, Y. J. 2006. Understanding electrodeposition of polyaniline coatings for corrosion prevention applications using the wire beam electrode method. *Corrosion Science*, **48**, 2274–2290.

12. Arenas, M., Bajos, L. G., De Damborenea, J. & Ocón, P. 2008. Synthesis and electrochemical evaluation of polypyrrole coatings electrodeposited onto AA-2024 alloy. *Progress in Organic Coatings*, **62**, 79–86.

13. Kousik, G., Pitchumani, S. & Renganathan, N. 2001. Electrochemical characterization of polythiophene-coated steel. *Progress in Organic Coatings*, **43**, 286–291.

14. Düdükcü, M. & Avcı, G. 2016. Electrochemical synthesis and corrosion inhibition performance of poly-5-aminoindole on stainless steel. *Progress in Organic Coatings*, **97**, 110–114.

15. Mousavi, Z., Alaviuhkola, T., Bobacka, J., Latonen, R. M., Pursiainen, J., & Ivaska, A. 2008. Electrochemical characterization of poly(3,4-ethylenedioxythiophene) (PEDOT) doped with sulfonated thiophenes. *Electrochimica Acta*, **53**, 3755–3762.

16. Shanmugham, C. & Rajendran, N. 2015. Corrosion resistance of poly p-phenylenediamine conducting polymer coated 316L SS bipolar plates for Proton Exchange Membrane Fuel Cells. *Progress in Organic Coatings*, **89**, 42–49.

17. Al Zoubi, M. & Endres, F. 2011. Electrochemical synthesis of poly(p-phenylene) and poly(p-phenylene)/TiO$_2$ nanowires in an ionic liquid. *Electrochimica Acta*, **56**, 5872–5876.

18. Duran, B., Çakmakcı, İ. & Bereket, G. 2013. Role of supporting electrolyte on the corrosion performance of poly(carbazole) films deposited on stainless steel. *Corrosion Science*, **77**, 194–201.

19. Lei, X. P., Liu, Z. Y., Wang, C., Qiang, F. U., Wang, B. X. & Laboratory, F. M. 2017. Influence of Curing Agent on Properties of Epoxy Resin Based Organic Attapulgite Clay/Polyaniline Coating. *Bulletin of the Chinese Ceramic Society*, **104**, 1001–1625.

20. Mahmoudian, M., Alias, Y. & Basirun, W. 2012. Effect of narrow diameter polyaniline nanotubes and nanofibers in polyvinyl butyral coating on corrosion protective performance of mild steel. *Progress in Organic Coatings*, **75**, 301–308.

21. Karmakar, H. S., Arukula, R., Thota, A., Narayan, R. & Rao, C. R. 2018. Polyaniline-grafted polyurethane coatings for corrosion protection of mild steel surfaces. *Journal of Applied Polymer Science*, **135**, 45806.

22. Laco, J. I. I., Villota, F. C. & Mestres, F. L. 2005. Corrosion protection of carbon steel with thermoplastic coatings and alkyd resins containing polyaniline as conductive polymer. *Progress in Organic Coatings*, **135**, 151–160.

23. Li, W. 2006. Study on polyaniline with acrylic resin by blending and preparation of compound conductive coatings. *New Chemical Materials*, **52**, 49–51.

24. Umoren, S. A. & Solomon, M. M. 2019. Protective polymeric films for industrial substrates: A critical review on past and recent applications with conducting polymers and polymer composites/nanocomposites. *Progress in Materials Science*, **104**, 380–450.

25. Ohtsuka, T., Iida, M. & Ueda, M. 2006. Polypyrrole coating doped by molybdo-phosphate anions for corrosion prevention of carbon steels. *Journal of Solid State Electrochemistry*, **10**, 714–720.

26. Wu, J. G., Chen, J. H., Liu, K. T. & Luo, S. C. 2019. Engineering Antifouling Conducting Polymers for Modern Biomedical Applications. *ACS Appl Mater Interfaces*, **11**, 21294–21307.

27. Contri, G., Barra, G. M. O., Ramoa, S. D. A. S., Merlini, C., Ecco, L. G., Souza, F. S. & Spinelli, A. 2018. Epoxy coating based on montmorillonite-polypyrrole: Electrical properties and prospective application on corrosion protection of steel. *Progress in Organic Coatings*, **114**, 201–207.

28. Palraj, S., Selvaraj, M., Vidhya, M. & Rajagopal, G. 2012. Synthesis and characterization of epoxy–silicone–polythiophene interpenetrating polymer network for corrosion protection of steel. *Progress in Organic Coatings*, **75**, 356–363.

29. Pourhashem, S., Saba, F., Duan, J., Rashidi, A., Guan, F., Nezhad, E. G. & Hou, B. 2020. Polymer/Inorganic nanocomposite coatings with superior corrosion protection performance: A review. *Journal of Industrial and Engineering Chemistry*, **88**, 29–57.

30. Chen, Z., Yang, W., Yin, X., Chen, Y., Liu, Y. & Xu, B. Corrosion protection of 304 stainless steel from a smart conducting polypyrrole coating doped with pH-sensitive molybdate-loaded $TiO_2$ nanocontainers. *Progress in Organic Coatings*, **146**, 105750.

31. Pagotto, J., Recio, F., Motheo, A. & Herrasti, P. 2016. Multilayers of PAni/n-$TiO_2$ and PAni on carbon steel and welded carbon steel for corrosion protection. *Surface and Coatings Technology*, **289**, 23–28.

32. Mostafaei, A. & Nasirpouri, F. 2014. Epoxy/polyaniline–ZnO nanorods hybrid nanocomposite coatings: Synthesis, characterization and corrosion protection performance of conducting paints. *Progress in Organic coatings*, **77**, 146–159.

33. Shi, S., Zhang, Z. & Yu, L. 2017. Hydrophobic polyaniline/modified $SiO_2$ coatings for anticorrosion protection. *Synthetic Metals*, **233**, 94–100.

34. Lei, Y. H., Qiu, Z. C. & Tan, N. *et al.* 2020. Polyaniline/$CeO_2$ nanocomposites as corrosion inhibitors for improving the corrosive performance of epoxy coating on carbon steel in 3.5% NaCl solution. *Progress in Organic Coatings*, **139**, 105430.

35. Kumar, A. M., Babu, R. S., Ramakrishna, S. & De Barros, A. L. 2017. Electrochemical synthesis and surface protection of polypyrrole-$CeO_2$ nanocomposite coatings on AA2024 alloy. *Synthetic Metals*, **234**, 18–28.

36. Izadi, M., Shahrabi, T. & Ramezanzadeh, B. 2018. Synthesis and characterization of an advanced layer-by-layer assembled $Fe_3O_4$/polyaniline

nanoreservoir filled with Nettle extract as a green corrosion protective system. *Journal of Industrial and Engineering Chemistry*, **57**, 263–274.

37. Ates, M. & Topkaya, E. 2015. Nanocomposite film formations of polyaniline via $TiO_2$, Ag, and Zn, and their corrosion protection properties. *Progress in Organic Coatings*, **82**, 33–40.

38. Krishnan, A., Joseph, B., Bhaskar, K. M., Suma, M. & Shibli, S. 2019. Unfolding the anticorrosive characteristics of $TiO_2$–$WO_3$ mixed oxide reinforced polyaniline composite coated mild steel in alkaline environment. *Polymer Composites*, **40**, 2400–2409.

39. Gao, F. J., Mu, J., Bi, Z. X., Wang, S. & Li, Z. L. 2021. Recent advances of polyaniline composites in anticorrosive coatings: A review. *Progress in Organic Coatings*. **151**, 106071

40. Gu, W., Zhang, H., Chen, C. & Zhang, J. 2022. Study on the design of ZnO/PANI composites and the mechanism of enhanced humidity sensing properties. *Current Applied Physics*, **34**, 112–121.

41. Althomali, R. H., Alamry, K. A., Hussein, M. A. & Guedes, R. M. 2022. Hybrid PANI@dialdehyde carboxymethyl cellulose/ZnO nanocomposite modified glassy carbon electrode as a highly sensitive electrochemical sensor. *Diamond and Related Materials*, **122**, 108803.

42. Belabed, C. et al. 2021. High photocatalytic performance for hydrogen production under visible light on the hetero-junction Pani-ZnO nanoparticles. *International Journal of Hydrogen Energy*, **46**, 17106–17115.

43. Sáaedi, A., Shabani, P. & Yousefi, R. 2019. High performance of methanol gas sensing of ZnO/PAni nanocomposites synthesized under different magnetic field. *Journal of Alloys and Compounds*, **802**, 335–344.

44. Najjar, R., Katourani, S. A. & Hosseini, M. G. 2018. Self-healing and corrosion protection performance of organic polysulfide@urea-formaldehyde resin core-shell nanoparticles in epoxy/PANI/ZnO nanocomposite coatings on anodized aluminum alloy. *Progress in Organic Coatings*, **124**, 110–121.

45. Hu, C., Li, Y., Kong, Y. & Ding, Y. 2016. Preparation of poly(o-toluidine)/nano ZnO/epoxy composite coating and evaluation of its corrosion resistance properties. *Synthetic Metals*, **214**, 62–70.

46. Sambyal, P., Ruhi, G., Mishra, M., Gupta, G. & Dhawan, S. K. 2018. Conducting polymer/bio-material composite coatings for corrosion protection. *Materials and Corrosion*, **69**, 402–417.

47. Lashkenari, M. S., Ghorbani, M., Silakhori, N. & Karimi-Maleh, H. 2021. Enhanced electrochemical performance and stability of Pt/Ni electrocatalyst supported on $SiO_2$-PANI nanocomposite: A combined experimental and theoretical study. *Materials Chemistry and Physics*, **262**, 124290.

48. Ding, J. & Cheng, L. 2021. Core-shell Fe3O4@$SiO_2$@PANI composite: Preparation, characterization, and applications in microwave absorption. *Journal of Alloys and Compounds*, **881**, 160574.

49. Shi, S., Zhao, Y., Zhang, Z. & Yu, L. 2019. Corrosion protection of a novel $SiO_2$@PANI coating for Q235 carbon steel. *Progress in Organic Coatings*, **132**, 227–234.

50. Toledo, R. P., Dias, C. E. S., Huanca, D. R. & Salcedo, W. J. 2018. Physical and chemical characterization of PANI/SiO$_2$/MPS heterostructure to be used as high sensitivity chemosensor for naphthalene. *Sensors and Actuators B: Chemical*, **277**, 445–455.

51. Wei, L., Chen, Q. & Gu, Y. 2010. Effects of inorganic acid in DBSA-PANI polymerization on transparent PANI-SiO$_2$ hybrid conducting films. *Journal of Alloys and Compounds*, **501**, 313–316.

52. Sumi, V. et al. 2020. PANI-Fe$_2$O$_3$ composite for enhancement of active life of alkyd resin coating for corrosion protection of steel. *Materials Chemistry and Physics*, **247**, 122881.

53. De León-Almazan, C. M., Estrada-Moreno, I. A., Páramo-García.U. & Rivera-Armenta, J. L. 2018. Polyaniline/clay nanocomposites. A comparative approach on the doping acid and the clay spacing technique, *Synthetic metals*, **236**, 61–67.

54. Akbarinezhad, E., Ebrahimi, M., Sharif, F. & Ghanbarzadeh, A. 2014. Evaluating protection performance of zinc rich epoxy paints modified with polyaniline and polyaniline-clay nanocomposite. *Progress in Organic Coatings*, **77**, 1299–1308.

55. Kalaivasan, N. & Syed Shafi, S. 2017. Enhancement of corrosion protection effect in mechanochemically synthesized Polyaniline/MMT clay nanocomposites. *Arabian Journal of Chemistry*, **10**, S127–S133.

56. Zhu, X. et al. 2019. In-situ modulation of interactions between polyaniline and graphene oxide films to develop waterborne epoxy anticorrosion coatings. *Progress in Organic Coatings*, **133**, 106–116.

57. Saeb, M. R. & Zarrintaj, P. 2019. In *Fundamentals and Emerging Applications of Polyaniline*, eds Mozafari, M. & Chauhan, N. P. S., 165–175. Elsevier.

58. Fazli-Shokouhi, S., Nasirpouri, F. & Khatamian, M. 2019. Polyaniline-modified graphene oxide nanocomposites in epoxy coatings for enhancing the anticorrosion and antifouling properties. *Journal of Coatings Technology and Research*, **16**, 983–997.

59. Ramezanzadeh, B., Moghadam, M. M., Shohani, N. & Mahdavian, M. 2017. Effects of highly crystalline and conductive polyaniline/graphene oxide composites on the corrosion protection performance of a zinc-rich epoxy coating. *Chemical Engineering Journal*, **320**, 363–375.

60. Liu, S., Pan, T., Wang, R., Yue, Y. & Shen, J. 2019. Anti-corrosion and conductivity of the electrodeposited graphene/polypyrrole composite coating for metallic bipolar plates. *Progress in Organic Coatings*, **136**, 105237.

61. Gao, Z., Yang, W., Yan, H., Yao, Y. & Ma, J. 2013. Electrochemical synthesis of layer-by-layer reduced graphene oxide sheets/polyaniline nanofibers composite and its electrochemical performance. *Electrochimica Acta*, **91**, 185–194.

62. Farag, A. A., Kabel, K. I., Elnaggar, E. M. & Al-Gamal, A. G. 2017. Influence of polyaniline/multiwalled carbon nanotube composites on

alkyd coatings against the corrosion of carbon steel alloy. *Corrosion Reviews*, **35**, 85–94.

63. Van, V. T. H. et al. 2018. Synthesis of silica/polypyrrole nanocomposites and application in corrosion protection of carbon steel. *Journal of Nanoscience and Nanotechnology*, **18**, 4189–4195.

64. Merisalu, M. *et al.* 2015. Graphene–polypyrrole thin hybrid corrosion resistant coatings for copper. *Synthetic metals*, **200**, 16–23.

65. Qiu, S., Li, W., Zheng, W., Zhao, H. & Wang, L. 2017. Synergistic effect of polypyrrole-intercalated graphene for enhanced corrosion protection of aqueous coating in 3.5% NaCl solution. *ACS applied materials & interfaces*, **9**, 34294–34304.

66. Alam, R., Mobin, M. & Aslam, J. 2016. Polypyrrole/graphene nanosheets/rare earth ions/dodecyl benzene sulfonic acid nanocomposite as a highly effective anticorrosive coating. *Surface and Coatings Technology*, **307**, 382–391.

67. Jiang, L., Syed, J. A., Lu, H. & Meng, X. 2019. In-situ electrodeposition of conductive polypyrrole-graphene oxide composite coating for corrosion protection of 304SS bipolar plates. *Journal of Alloys and Compounds*, **770**, 35–47.

68. Tehrani, M. E. H. N., Ramezanzadeh, M., Bahlakeh, G. & Ramezanzadeh, B. 2021. S, P-codoped rGO-phytic acid-polythiophene core–shell; synthesis, modeling, and dual active–passive anti-corrosion performance of epoxy nanocomposite. *Journal of Industrial and Engineering Chemistry*, **103**, 102–117.

69. Prabakar, S. R. & Pyo, M. 2012. Corrosion protection of aluminum in $LiPF_6$ by poly (3, 4-ethylenedioxythiophene) nanosphere-coated multi-walled carbon nanotube. *Corrosion science*, **57**, 42–48.

70. Zhu, Q. *et al.* 2020. Synergistic effect of polypyrrole functionalized graphene oxide and zinc phosphate for enhanced anticorrosion performance of epoxy coatings. *Composites Part A: Applied Science and Manufacturing*, **130**, 105752.

71. Fekri, F., Shahidi Zandi, M. & Foroughi, M. M. 2019. Polypyrrole coatings for corrosion protection of Al alloy2024: influence of electrodeposition methods, solvents, and ZnO nanoparticle concentrations. *Iranian Polymer Journal*, **28**, 577–585.

72. Jadhav, N., Kasisomayajula, S. & Gelling, V. J. 2020. Polypyrrole/metal oxides-based composites/nanocomposites for corrosion protection. *Frontiers in Materials*, **7**, 95.

73. Deshpande, P. P., Jadhav, N. G., Gelling, V. J. & Sazou, D. 2014. Conducting polymers for corrosion protection: a review. *Journal of Coatings Technology and Research*, **11**, 473–494.

74. Ren, Y. & Zeng, C. 2008. Effect of conducting composite polypyrrole/polyaniline coatings on the corrosion resistance of type 304 stainless steel for bipolar plates of proton-exchange membrane fuel cells. *Journal of Power Sources*, **182**, 524–530.

75. Paliwoda-Porebska, G. *et al.* 2005. On the development of polypyrrole coatings with self-healing properties for iron corrosion protection. *Corrosion science*, **47**, 3216–3233.

76. Kinlen, P. J., Menon, V. & Ding, Y. 1999. A mechanistic investigation of polyaniline corrosion protection using the scanning reference electrode technique. *Journal of the Electrochemical Society*, **146**, 3690.

77. Hao, Y., Sani, L. A., Ge, T. & Fang, Q. 2017. Phytic acid doped polyaniline containing epoxy coatings for corrosion protection of Q235 carbon steel. *Applied Surface Science*, **419**, 826–837.

78. Pour-Ali, S., Dehghanian, C. & Kosari, A. 2014. In situ synthesis of polyaniline–camphorsulfonate particles in an epoxy matrix for corrosion protection of mild steel in NaCl solution. *Corrosion Science*, **85**, 204–214.

79. Kowalski, D., Ueda, M. & Ohtsuka, T. 2010. Self-healing ion-permselective conducting polymer coating. *Journal of Materials Chemistry*, **20**, 7630–7633.

80. Deyab, M. 2014. Corrosion protection of aluminum bipolar plates with polyaniline coating containing carbon nanotubes in acidic medium inside the polymer electrolyte membrane fuel cell. *Journal of Power Sources*, **268**, 50–55.

81. Chang, C. H. *et al.* 2012. Novel anticorrosion coatings prepared from polyaniline/graphene composite. *Carbon*, **50**, 5044–5051.

82. Wessling, B. 1998. Dispersion as the link between basic research and commercial applications of conductive polymers (polyaniline). *Synthetic Metals*, **93**, 143–154.

83. Marchebois, H., Keddam, M., Savall, C., Bernard, J. & Touzain, S. 2004. Zinc-rich powder coatings characterisation in artificial sea water: EIS analysis of the galvanic action. *Electrochimica Acta*, **49**, 1719–1729.

84. Meroufel, A. & Touzain, S. 2007. EIS characterisation of new zinc-rich powder coatings. *Progress in Organic Coatings*, **59**, 197–205.

85. Shen, S. & Zuo, Y. 2014. The improved performance of Mg-rich epoxy primer on AZ91D magnesium alloy by addition of ZnO. *Corrosion Science*, **87**, 167–178.

86. Liu, J. *et al.* 2018. Insight into the impact of conducting polyaniline/graphene nanosheets on corrosion mechanism of zinc-rich epoxy primers on low alloy DH32 steel in artificial sea water. *Journal of The Electrochemical Society*, **165**, C878.

87. Schaefer, K. & Miszczyk, A. 2013. Improvement of electrochemical action of zinc-rich paints by addition of nanoparticulate zinc. *Corrosion Science*, **66**, 380–391.

88. Chen, W. B., Chen, P., Chen, H. Y., Wu, J. & Tsai, W. T. 2002. Development of Al-containing zinc-rich paints for corrosion resistance. *Applied Surface Science*, **187**, 154–164.

89. Gergely, A., Pászti, Z., Mihály, J., Drotár, E. & Török, T. 2015. Galvanic function of zinc-rich coatings facilitated by percolating structure of the

carbon nanotubes. Part I: Characterization of the nano-size particles. *Progress in Organic Coatings*, **78**, 437–445.

90. Hayatdavoudi, H. & Rahsepar, M. 2017. A mechanistic study of the enhanced cathodic protection performance of graphene-reinforced zinc rich nanocomposite coating for corrosion protection of carbon steel substrate. *Journal of Alloys and Compounds*, **727**, 1148–1156.

91. Meroufel, A., Deslouis, C. & Touzain, S. 2008. Electrochemical and anticorrosion performances of zinc-rich and polyaniline powder coatings. *Electrochimica Acta*, **53**, 2331–2338.

92. Tian, Z., et al. 2014. Recent progress in the preparation of polyaniline nanostructures and their applications in anticorrosive coatings. *Rsc Advances*, **4**, 28195–28208.

93. Wei, C., Wang, J., Quan, X., Song, Z. & Zhi, W. 2018. Antifouling and anticorrosion properties of one-pot synthesized dedoped bromo-substituted polyaniline and its composite coatings. *Surface and Coatings Technology*, **334**, 7–18.

94. Digital Journal. 2022. Waterborne Coatings Market Evolving Technology, Trends and Industry Analysis:2027. *Digital Journal*. https://www.digitaljournal.com/pr/waterborne-coatings-market-evolving-technology-trends-and-industry-analysis-2027#ixzz7SgBMWNT2.

95. Gurunathan, T., Arukula, R., Chung, J. S. & Rao, C. 2016. Development of environmental friendly castor oil-based waterborne polyurethane dispersions with polyaniline. *Polymers for Advanced Technologies*, **27**, 1535–1540.

96. Tundo, P. et al. 2000. Synthetic pathways and processes in green chemistry. Introductory overview. *Pure and Applied Chemistry*, **72**, 1207–1228.

97. Wang, H. et al. 2017. Facile approach to fabricate waterborne polyaniline nanocomposites with environmental benignity and high physical properties. *Scientific Reports*, **7**, 43694.

98. Wu, X., Lu, C., Xu, H., Zhang, X. & Zhou, Z. 2014. Biotemplate Synthesis of Polyaniline@Cellulose Nanowhiskers/Natural Rubber Nanocomposites with 3D Hierarchical Multiscale Structure and Improved Electrical Conductivity. *ACS Applied Materials & Interfaces*, **6**, 21078–21085.

99. Zhu. X., Ni. Z., Dong. L., Yang. Z., Chen. L. & Zhou. X. 2019, In-situ modulation of interactions between polyaniline and graphene oxide films to develop waterborne epoxy anticorrosion coatings, *Progress in organic coatings*, **133**, 106–116.

100. Morsi, R. E., Khamis, E. A. & Al-Sabagh, A. M. 2016. Polyaniline nanotubes: Facile synthesis, electrochemical, quantum chemical characteristics and corrosion inhibition efficiency. *Journal of the Taiwan Institute of Chemical Engineers*, **60**, 573–581.

101. Lee, W. P., Brenneman, K. R., Hsu, C. H., Shih, H. & Epstein, A. J. 2001. Charge Transport Properties of High-Strength, High-Modulus Sulfonated

Polyaniline/Poly(p-phenylene terephthalamide) Fibers. *Macromolecules*, **34**, 2648–2652.

102. Chan, H., Ho, P., Ng, S. C., Tan, B. & Tan, K. L. 1995. A New Water-Soluble, Self-Doping Conducting Polyaniline from Poly(o-aminobenzylphosphonic acid) and Its Sodium Salts: Synthesis and Characterization. *Journal of the American Chemical Society*, **117**, 8517–8523.

103. Zhang, H. M. & Wang, X. H. 2013. Eco-friendly water-borne conducting polyaniline. *Polymer Science: English edition*, **31**, 17.

104. Hua, F. & Ruckenstein, E. 2004. Copolymers of Aniline and 3-Aminophenol Derivatives with Oligo(oxyethylene) Side Chains as Novel Water-Soluble Conducting Polymers. *Macromolecules*, **37**, 6104–6112.

105. Swapna, P., Rao, S. S. & Sathyanarayana, D. N. 2010. Water-soluble conductive blends of polyaniline and poly(vinyl alcohol) synthesized by two emulsion pathways. *Journal of Applied Polymer ence*, **98**, 583–590.

106. Li, W. *et al.* 2002. Toward Understanding and Optimizing the Template-Guided Synthesis of Chiral Polyaniline Nanocomposites. *Macromolecules*, **35**, 9975–9982.

107. Wang, Y. *et al.* 2010. Morphological study on water-borne conducting polyaniline-poly(ethylene oxide) blends. *Journal of Polymer Science Part B Polymer Physics*, **40**, 605–612.

108. Zhu, K., Li, J., Wang, H. & Fei, G. 2020. Comparative study on anticorrosion enhancement of carboxylated and sulfonated self-doped polyaniline on waterborne epoxy coating. *Journal of Macromolecular Science Part A*, **58**, 1–13.

109. Qiu, S. H., Chen, C., Cui, M. J., Zhao, H. C. & Wang., L. P. 2017. Corrosion protection performance of waterborne epoxy coatings containing self-doped polyaniline nanofiber. *Applied Surface Science*, **407**, 213–222.

110. Cai, K. *et al.* 2016. Preparation of polyaniline/graphene composites with excellent anti-corrosion properties and their application in waterborne polyurethane anticorrosive coatings. *RSC Advances*, **6**, 95965–95972.

111. Xiao, F. *et al.* 2018. Anticorrosive durability of zinc-based waterborne coatings enhanced by highly dispersed and conductive polyaniline/graphene oxide composite. *Progress in Organic Coatings*, **125**, 79–88.

112. Jeon, U. S., Hong, S. A., Oh, I. H. & Kang, S. G. 2005. Performance of a 1 kW-class PEMFC stack using TiN-coated 316 stainless steel bipolar plates. *Journal of Power Sources*, **142**, 177–183.

113. Huang, N., Liang, C. & Yi, B. 2008. Corrosion resistance of PANi-coated steel in simulated PEMFC anodic environment. *Materials and Corrosion*, **59**, 21–24.

114. Zhang, K. & Sharma, S. 2016. Site –selective, low-loading, Au nanoparticle-polyaniline hybrid coatings with enhanced corrosion resistance and conductivity for Fuel Cells. *Acs Sustainable C*, **5**, 277–286.

# An Introduction of Conducting Polymers in Anti-Bacterial and Anti-Biofouling Applications

Yanhua Lei and Tao Liu

## CONTENTS

DOI: 10.1201/9781003376194-6

## 6.1 INTRODUCTION TO MICROBIALLY INFLUENCED CORROSION AND MARINE BIOFOULING

Microbial fouling mainly refers to the phenomenon that bacteria and other microorganisms attach to the surface of materials and form a biofilm. In daily life, microbial contamination will accelerate food contamination and corruption, water pollution, and medical equipment pollution, serious threats to human physical and mental health. Bacterial infections accounted for about 33% of the 52 million deaths worldwide in 1995, according to the World Health Organization (WHO), and further the number is continuously growing. In the marine environment, microbial contamination will not only accelerate the corrosion of marine engineering materials, resulting in microbial corrosion, which usually refers to microbially influenced corrosion (MIC), but also accelerate the attachment of large organisms on the surface of marine engineering materials, resulting in serious contamination of large organisms. MIC is often produced by a mixture of anaerobic sulfate-reducing bacteria (SRB) and aerobic iron-oxidizing bacteria (IOB).[1,2] During this process, SRB and IOB cooperate together to form biofilms on metal surfaces that are usually composed of sessile cells, extracellular polymeric substances (EPS), and corrosion products from these two bacteria and play a very important role in MIC.[3,4] A biofilm layer creates conditions on a steel substrate that accelerates corrosion and facilitates the subsequent attachment of marine organisms.

There are about 2,000–3,000 species of marine fouling organisms in the ocean, and about 50–100 species of common marine fouling organisms, including sessile organisms, microorganisms, and plants. Marine biofouling comes from the undesirable settlement and accumulation of marine microorganisms, plants, and animals on submerged surfaces of materials and it has a huge adverse influence on the infrastructure and equipment served in marine industries.[5]

Marine biofouling will cause great harm to ship and marine facilities, usually leading to the destruction of metal surface coating, accelerating metal corrosion, and increasing ship navigation resistance. Marine fouling generates surface roughness, which increases the drag resistance of a ship moving through water and consequently increases fuel consumption and emission of greenhouse gases. Heavy calcareous fouling may result in powering penalties of more than 85%. Moreover, even slime films can lead to significant increases in resistance and powering (approximately 20%).[6]

Another effect of marine fouling is the deterioration of coatings, such as favored corrosion. Settlement of fouling generates an increase of the frequency of dry-docking operations either because of the need of additional hull cleaning or even in costly additional coating replacement or hull repair (see Figures 6.1–6.2).

According to incomplete statistics, biological contamination of marine engineering materials causes a direct economic loss of nearly 1 trillion RMB and energy waste of more than 30% of the vehicles every year, which has become one of the technical bottlenecks that seriously restrict the development of marine engineering technology and equipment. It has been reported that the U.S. Navy alone costs about $1 billion a year to deal with marine fouling. In addition, marine fouling can cause serious environmental and health impacts. Fouled vessels are the most common vectors of marine species that attach themselves to the ship hull and can be displaced in foreign areas, leading to the introduction of invasive, non-indigenous species into non-native environments.

FIGURE 6.1   Pictures of defiled marine biofouling cases.

Source: created by author for this publication.

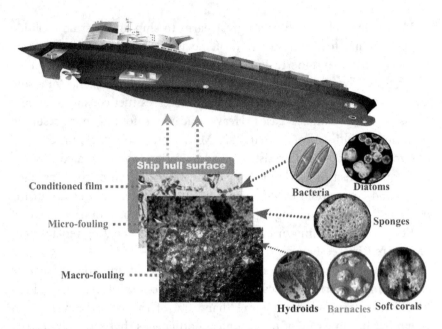

FIGURE 6.2 A schematic illustration of the three common fouling steps, including molecular (conditioned film) and micro- and macro-fouling.

**Source: Progress in Materials Science (Elsevier).**

## 6.2 HOW DOES MARINE LIFE ATTACH TO THE SURFACE OF MATERIALS?

In fact, the attachment of marine organisms goes through a complex process. As shown in Figure 6.3A, at the initial stage, small organic molecules are adsorbed on the surface of the material to form the base membrane. At the same time, common bacteria in seawater, such as *Helicobacter, Vibrio, Coccus,* and *Bacillus genus,* also attach to the surface of the material, and they construct a biofilm in a few days to lay a foundation for subsequent biological attachment. In the middle stage, diatom spores, ciliates, bell worms, nematodes, and other small fouling organisms are adsorbed on the biofilm, and the number increases. In the later stage, some attached algae are connected with the surface of the object by secreting glia, resulting in a sharp increase in the species and number. The establishment of biofilm laid a foundation for the attachment of some large marine attached organisms such as mussels and barnacles to the surface of the material. Therefore, inhibition of biofilm formation is often used to effectively control the problem of biological fouling.

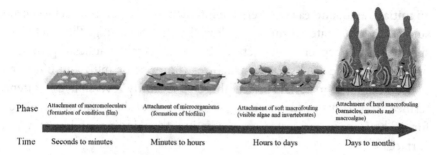

FIGURE 6.3A   Schematic diagram of marine biological contamination.

**Source: Science of The Total Environment (Elsevier).**

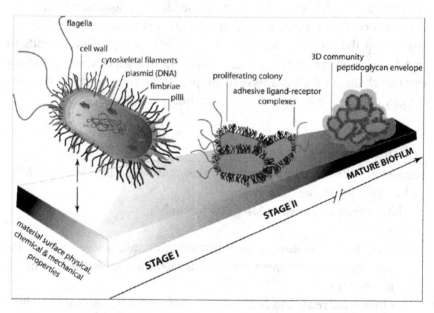

FIGURE 6.3B   Schematic of prokaryotic, prototypical bacterial cell structure and the two-stage bacterial adhesion model that precedes organization of a mature biofilm.

**Source: Macromolecules (American Chemical Society).**

It is clear that adhesion of viable bacteria to material surfaces is a necessary condition for biofilm formation in both hygienic and industrial contexts. However, this adhesion process is sufficiently complex that engineered prevention and promotion of bacterial adhesion remains an elusive goal. As outlined in Figure 6.3B, the basic stages of bacterial adhesion are generally described by a two-stage kinetic binding model[7]:

an initial, rapid, and easily reversible interaction between the bacteria cell surface and the material surface, followed by a second stage that includes specific and nonspecific interactions between so-called adhesin proteins expressed on bacterial surface structures (fimbriae or pilli) and binding molecules on the material surface; this step is slowly reversible and often termed *irreversible*. And various types of long- and short-range attractive forces have been suggested as mediating the binding of bacteria to surfaces.

A biofilm layer creates conditions on a steel substrate that accelerates corrosion and facilitates the subsequent attachment of marine organisms. Fouling organisms are attached to the steel plate, due to the role of sulfate reducing bacteria, iron bacteria, so that the aquaculture cage on the metal corrosion intensified. Some fouling organisms will destroy the coating on the metal surface, so that the metal is exposed and leads to metal corrosion. The fouling organisms with lime shells cover the metal surface, changing the local oxygen supply of the metal surface, forming oxygen concentration cells and accelerating corrosion.

## 6.3 COMPARISON BETWEEN MIC AND MARINE BIOFOULING

Ning et al.[1] summarized the differences and similarities between MIC and marine biofouling as follows:

The differences between MIC and biofouling:

1. MIC is a corrosion process occurring at micro level, while bio-fouling is a settlement and accumulation process occurring at macro level;
2. Organisms related to MIC are only different microorganisms, while organisms related to biofouling include different microorganisms, plants, and animals;
3. MIC damages the materials directly, while biofouling's damage is more wide and complicated in different areas.

The similarities between MIC and biofouling include the following points:

1. Both MIC and biofouling start from the formation of an absorbed film on the material surface;

2. The formations of MIC and biofouling are both closely related to the biofilms created by marine microorganisms;

3. Similar origins of MIC and biofouling lead to their similar prevention strategies. If a method could destroy formed biofilms and prevent the formation of new biofilms, it could eliminate or prevent MIC and biofouling at the same time.

## 6.4 INTRODUCTION TO ANTI-FOULING COATINGS AND ANTI-BACTERIAL COATINGS

### 6.4.1 Introduction to Anti-Fouling Coatings

Anti-fouling is the process of removing and/or preventing the accumulation of this biofouling/fouling. There are a variety of ways to protect an underwater surface from fouling, which is commonly used in anti-foul coatings.

Anti-fouling coatings are a mixture of non-toxic and toxic chemicals that provide a buffer between the submerged surface and the marine environment. This coating, when containing tributyltin moiety (TBT), has been known to cause defective shell growth in some oysters and other organic organisms, even when in extremely low concentrations. In some countries, the TBT substance has been banned from use in the marine environment. By limiting the adhesion of marine organisms, it is estimated that marine anti-fouling coatings provide the shipping industry with annual fuel savings of $60 billion and reduced emissions of 384 million and 3.6 million tons, respectively, for carbon dioxide and sulfur dioxide per annum.[8]

With the increasing concern to the marine ecological environment, it is therefore urgent to choose other anti-fouling additives that could be effective, but not harmful to the environment. TBT-free formulations, in which tin is replaced by metals like copper, zinc, and titanium, are in use today. Currently, cuprous oxide as anti-fouling agent is extensively used in marine coatings. Although copper is essential for all forms of life due to cellular processes, it is considered toxic when tolerance limits are exceeded. Copper has excellent anti-fouling properties against barnacles and algae, even though some species are resistant to this metal. Additionally, one of the most common zinc-based biocides, zinc pyrithione (ZnPT), was found to be toxic to several marine creatures. The aforementioned drawbacks are inevitable for metal-based biocides, because they derive from the undegradable nature of metal ions in environments. Therefore, the development of green, environmental protection, anti-microbial adhesion, and excellent anti-fouling agent has become an important research topic in the field of science, especially in the field of marine science.

Currently, another major kind of anti-fouling coating termed "fouling release coatings", has a low surface tension and low elastic modulus, thus achieving a low energy surface.[9] Fouling organisms can hardly adhere to fouling release surfaces, and the attached organisms can also be easily removed because of the weak adhesion. There are no biocides released from the coatings to kill the absorbed organisms, and thus, those coatings are relative moderate to the environment. Silicone or fluorine-based polymers, such as polydimethylsiloxane (PDMS) and Teflon (PTFE) were mainly applied to construct these coatings.[10]

In recent years, some organic acids, inorganic acids, such as, lactones, terpenoids, phenols, sterols and indoles extracted from animals and plants have good anti-fouling properties and have been used as anti-fouling agents in anti-fouling coatings with the characteristics of green environmental protection, high efficiency, and ability to last. However, due to the low content of anti-fouling components in most plants and animals, cumbersome extraction process, and high cost, the natural product anti-fouling agents have not been widely used, and there are still many problems to be solved urgently.

In nature, there is some biology with a special natural surface, who use the "natural" surface characteristics to resist the attachment of other substances, and the inspiration of a bionic anti-fouling coating is derived from this. For example, the lotus leaf in the pond, dolphins, whales, sharks, algae, and so on in the ocean, their surface has a special structure, which endows them with the characteristics of low surface energy, and thus can effectively prevent the attachment of organisms. Inspired by the anti-fouling mechanism of these organisms, bionic anti-fouling coatings with similar characteristics have been developed. It is believed that biomimetic anti-fouling strategies will contribute to the development of nontoxic anti-fouling techniques with exceptional repellency and stability. Several researchers summarized biomimetic nanocomposite coatings for marine anti-fouling with good fouling release performance.[9–12] For example, Chen et al.[9] reviewed three basic bioinspired strategies, including natural anti-foulant coatings, bioinspired polymeric coatings, and biomimetic surface topography coatings, as shown in Figure 6.4. However, most of the

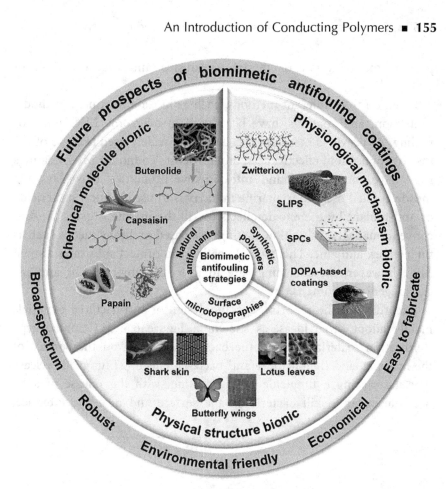

FIGURE 6.4  The three bioinspired polymeric coatings for marine anti-fouling.

**Source: Science of The Total Environment (Elsevier).**

bionic anti-fouling coatings are in the stage of laboratory research, and it will take some time for further industrialization.

## 6.4.2 Introduction to Anti-Bacterial Coatings

As mentioned in Figure 6.3, biofouling usually begins with the initial bacterial attachment followed with the formation of biofilms. According to the studies of mechanisms of MIC and marine biofouling, biofilm is the main triggering condition for both of them. Thus, to prevent marine fouling, researchers often need to start with holding back the attachment of bacteria. As a result, to prevent and eliminate MIC or marine bio-fouling on the surface of materials served in marine environment, it is important to control the activity of microorganisms in biofilms or to prevent the adhesion of marine organisms and the formation of biofilms[1].

Based on this, different methods, such as biocide treatment, cathodic protection, and coatings, have been used to overcome MIC and bio-fouling. As previous mentioned, coatings are widely used because of the ease of application, effectiveness, and low cost. To inhibit MIC, coatings must have anti-bacterial and anti-corrosive properties. Nowadays, the mostly applied method in marine industry is to use chemical bactericide or anti-foulant by their toxic effects to kill the marine organisms. And conventional MIC-inhibition coatings are based on heavy metals such as tin, copper, and zinc. This type of coating can protect substrates against MIC; however, these coatings are toxic to the environment and are carcinogenic to humans.

According to the reports,[1,7,13] the design of anti-bacterial materials mainly adopt three ideas, as shown in Figure 6.5: (1) Contact-type sterilization material can kill the bacteria adsorbed on the surface of the material, and then achieve anti-bacterial effect; (2) fungicide release type, by placing a fungicide in the coating, with the release of anti-bacterial agent to kill bacteria on the surface and near the interface;

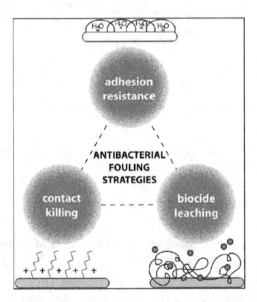

FIGURE 6.5   The design of anti-bacterial and anti-fouling materials mainly adopts three modes (contact sterilization material, release type, and anti-adhesion).

**Source: Macromolecules (American Chemical Society).**

(3) anti-adhesion anti-fouling material, the use of materials with anti-fouling ability, so that bacteria and pollutants are difficult to adsorb on the surface of the material, to achieve the effect of anti-bacterial. The ideal anti-microbial material would kill bacteria that cling to the surface and then release dead bacteria and other contaminants that cling to the surface. Therefore, the preparation of anti-microbial material system with multiple anti-microbial mechanisms is more important in damage prevention.

## 6.5 ANTI-BIOFOULING OR ANTI-BACTERIAL PERFORMANCE OF CONDUCTING POLYMERS

Conducting polymers (CPs) such as polyaniline (PANI), polypyrrole (PPy), and polythiophene and their derivatives and composites are attractive biomaterials due to their biocompatibility, facile synthesis and simple modification, and their ability to electronically control a range of physical and chemical properties by (i) surface functionalization techniques and (ii) the use of a wide range of molecules that can be entrapped or used as dopants.[14,15] Such CPs synthesized by chemical or electrochemical, have great applications in the area of electronics,[16] inkjet printers,[17] rechargeable batteries,[18] biosensor,[19] and biology.[15,20] Since the first application of PANI in anodic protection of steel, CPs have been employed in the manufacture of protective coatings.

CPs have broad application prospects in marine anti-bacterial and anti-fouling.[21] For example, in the 1990s, Mitsubishi Heavy Industries Co., Ltd. developed a new type of anti-fouling coating, which was composed of an inner insulation layer to insulate the direct contact between the hull substrate and the water of the sea, and of the CP-contained conductive layer outside. When small anodic current is applied to the outside conductive layer, then the seawater electrolysis occurs on the surface of the coating, generating hypochlorous acid and ion cover outside the coating surface, which results in the attachment of algae, shellfish, and other organisms. Due to the small current, the concentration of hypochlorite in seawater is far less than that of direct seawater electrolysis, and there is almost no negative effect on the marine environment. Therefore, it is of great significance to develop a sustainable anti-fouling conductive coating. However, this method is technically difficult, and the resin used, filler, curing agent, etc., need to be improved to prolong the coating life and anti-fouling effect. An important direction of this technology is to improve the conductivity and electrolytic resistance of coating.

Besides, the addition of PANI, PPy, and other conducting materials in the coatings can not only improve the anti-corrosion performance of the coatings, but also give the certain anti-fouling properties to the coatings. Paints containing cuprous oxide and PANI, or its sulphonated derivative (SPAN), showed a much more effective anti-fouling protection than paint containing only cuprous oxide, suggesting a synergistic effect between these two compounds. The PANI used as additives in anti-fouling paints could improve the efficiency of anti-fouling coatings, reduce the amount of $Cu_2O$ used in the formulations and hence its release to the marine environment.[22]

Among the various CPs, PANI has exhibited excellent anti-corrosive properties and been widely used in corrosion protection. Thus, in the following section, we mainly focus on PANI to illustrate the applications of CPs in anti-bacterial and anti-fouling.

## 6.5.1 Anti-Bacterial Properties of PANI

The anti-bacterial activity of PANI is due to their special structure that comprises amino and hydrocarbon groups, as well as the doping state and electro-redox activity.

N-Halamines represent an important subcategory of anti-bacterial agents that were extensively studied in recent years. A specific feature of this series of compounds is that they contain both nitrogen and halogen atoms that provide excellent anti-bacterial properties together. It is important to note that doped PANI and its derivatives also have similar moieties in their structures, which probably explains their high efficiency in bacterial growth inhibition. Thus, some CPs such as PANI, PPy, and polythiophene and their derivatives and composites are suitable for MIC inhibition because of the anti-corrosive and anti-bacterial properties. The high redox properties of conductive polymers can passivate metal, generating a protective oxide layer. Due to the positively charged nitro-groups, conductive polymers display biocidal properties inhibiting bacterial attachment and biofilm formation. In addition to the above factors, the anti-bacterial performance of CPs is closely related to their doping level, the conductivity as well as doped ions.

C. Dhivya et al.[23] synthesized the PANI doped with 2,4,6-trinitrophenol (picric acid, 3,5-dinitrobenzoic acid and hydrochloric acid) in the form of emerald salt by chemical oxidative polymerization. Polyaniline Emeraldine base (PANI-EB) was prepared by dedoping polyaniline chloride (PANI-Cl). The in vitro anti-bacterial properties of

PANI against gram-negative bacteria, gram-positive bacteria, and *Candida albicans* were evaluated by the Agar diffusion method. The anti-bacterial results of the diameter of bacteriostatic zone and minimum bacteriostatic concentration show that compared with PANI-EB, the doped PANI has an enhanced anti-bacterial effect. Similar results were obtained from the anti-bacterial investigation of the PANI doped with citric acid or rosin acid, which exhibited strong anti-bacterial activity against selected gram-positive bacteria and gram-negative bacteria, and the doped PANIs had higher anti-bacterial performance than PANI-EB.[24]

In the study,[25] electrically conductive membranes were prepared by coating graphene (Gr) and PANI doped with phytic acid (PA) on polyester filter cloth. The Gr/PANI-PA modified membrane had a good conductivity and an excellent anti-fouling property and its electric resistance was 93% lower than the PANI-HA (doped with HCl) modified membrane and 13% lower than the PANI-PA modified membrane. The anti-fouling performances of the modified conductive membranes were compared in EMBR (MBR, membrane bioreactor attached with a 0.2 V/cm electric field), tested by a short time filtration of a yeast suspension solution. The result indicated that the membrane with a higher conductivity had better anti-fouling property, smaller slopes in its t/V-V curves (adopting the classical cake filtration model, a suitable description for the membrane fouling process). The short-term cumulative permeate volume from the Gr/PANI-PA modified membrane was increased by ~ 58% after applying 0.2 V/cm electric field, while it was only ~ 10% for a PANI-PA modified membrane.

L.A. Gallaratoa et al.[26] covers the surface of the polymer (polyethylene terephthalate) with a PANI film by in-situ polymerization and micro-structures by direct laser interference pattern (DLIP). The live and dead bacterial cells were subsequently conducted to visually demonstrate cell integrity disruption on the different surfaces cultivated with P. aeruginosa as model bacteria for 48 hours at 37°C, in Figure 6.6. The number of live cells on PET films, 80 ± 1.5%, is higher than the dead cells (Figure 6.6(a–c)). The results of this study agree with the percentage of the adhered cells, the quantitation of the cell in the images shows an adhesion reduction of 50% and 60% for PET-PANI and PET-PANI-M. Therefore, it is possible to conclude that bacterial viability and adhesion are reduced on the PET-PANI and strongly reduced on PET-PANI-M compared to the PET surface.

Incorporating functional groups in the aromatic ring in PAIN molecular results in PANI derivatives that would affect the anti-bacterial

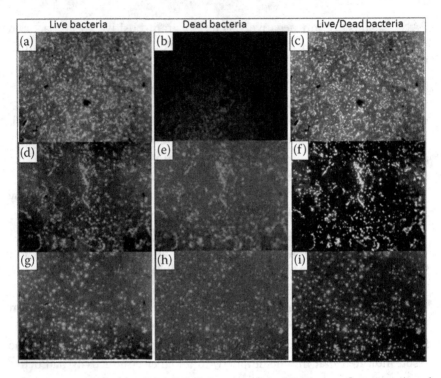

FIGURE 6.6 Epifluorescence digital images of the live (green fluorescent) and dead (red fluorescent) cells grown for 48 hours on (a–c) PET, (d–f) PET-PANI, and (g–i) PET-PANI-M.

**Source: Colloids and Surfaces B: Biointerfaces (Elsevier).**

performance of the polymers. Meanwhile, the PANI derivatives usually exhibited improved the solubility in water or organic solvents, which would greatly facilitate the application in the coating based on it. Recently, a detailed study on the effect of various functional groups both at the ortho position of the aromatic ring and in the amino group of PANI on the anti-bacterial properties of PANI against gram-positive (*B. subtilis*) and gram-negative (*P. aureofaciens*) microorganisms were investigated by Andriianova et al.[27] It was also found that the changes in the anti-bacterial activity result from a change in the nature of the nitrogen atom in PANI derivatives. N-substituted PANI derivatives demonstrated not only bactericidal but also bacteriostatic properties toward the test microorganisms. Varying the nature and position of the substituent allowed people to synthesize various N-substituted PANI derivatives with a high degree of doping, which is the most promising approach to PANI modification for application in bacterial growth inhibition.

Incorporation of the halogens in the PANI molecular would also be beneficial to improve the anti-bacterial performance of the polymer. Cai et al.[28] prepared brominated polyaniline (Br-PANI) by a new synthesis method using brominated polyaniline (Br-PANI) as raw material and potassium bromate and potassium bromide as bromate reagents. The experimental results indicate that both doped and dedoped Br-PANIs have long-term anti-bacterial activities as compared to doped/dedoped PANI. What's more, the anti-bacterial abilities increase with increasing the Br/N molar ratio of the Br-PANI. Figure 6.7 shows the growth of *E. coli* after contacting with soak solutions of different polymer samples. The soak solutions without polymer samples were obtained by soaking the polymer samples with NaCl solutions, then removing the soaked polymer samples and only keeping the solutions. As can be seen in Figure 6.7, the *E. coli* present different growth situations after contacting with different soak solutions for 2 hours. The soak solutions of doped PANI (S1) and doped Br- PANI (S2) show weak anti-bacterial abilities and the reduction of bacterial colonies reduced by 45% and 50%, respectively, while the soak solutions of dedoped PANI (T1) and dedoped Br-PANI (T2) have no anti-bacterial abilities compared to a blank group.

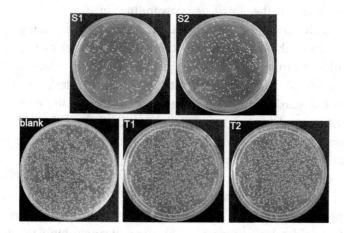

FIGURE 6.7   The growth of *E. coli* after contacting with soak solutions of different polymer samples.

Source: *Journal of Applied Polymer Science* (John Wiley & Sons Books).

The anti-bacterial activity of PANI is due to their special structure that comprises amino and hydrocarbon groups. PANI chains are positive in the salt form and neutral in the base form. Protonated PANI molecules carry multiple positive charges, while the outer shells of bacterial cells carry a negative charge.[29] The interaction between the polycations in PANI and cell membranes result in the formation of surface complexes between the membrane and the polycation stabilized by multiple ionic contacts of polycationic ammonium units with negatively charged groups of lipid molecules and membrane proteins. A PANI matrix binds to the cell walls through electrostatic interactions and gets anchored to the cell walls at several sites on the membranes. The ion channels present on the cell wall experience a change in the potential gradient because of which the normal influx and efflux of electrolytes are disturbed. It is well established that PANI materials are highly electroactive at low pH values. Thus, PANIs might bring about a change in the pH levels of the cell due to the release of the dopants. The permeability of the cell membranes on the bacterial cells is affected. The released dopant ions from the PANI matrix penetrate the semipermeable membranes and disturb the functions of the iongates. The dopant ions react with enzymes, amino acids, nucleic acids, DNA, and electrolytes present in the cell sap and may alter the cell composition, leading to cell lysis.[29]

It needs to be pointed out that the anti-bacterial properties of the PANI coating continue to degrade with the gradual reduction of the doped PANI to the intrinsic polyaniline in the soaking process. Therefore, the method of directly adding PANI to the coating still faces the problem of the degradation of the anti-bacterial properties.

In order to further expand its application, currently common solutions are to composite modification with other anti-bacterial materials to improve the stability and anti-bacterial properties of the coating. Anti-microbial materials such as nanometals,[30,31] ZnO,[32,33] $TiO_2$[34], $CeO_2$[35], Gr[36] and quaternary ammonium salt[15] were used to enhance the anti-bacterial properties of the polymer coatings.

For example, silver NPs (Ag NPs) have been known as effective anti-bacterial agents against a broad spectrum of bacteria. The anti-bacterial mechanism of Ag NPs has been well studied.[37] Firstly, the dissolution of Ag NPs would release silver ions that can strongly bind to important cellular components and efficiently inhibit vital cellular functions. Secondly, in the presence of oxygen, Ag NPs can generate reactive oxygen species, which are highly reactive and can strongly damage

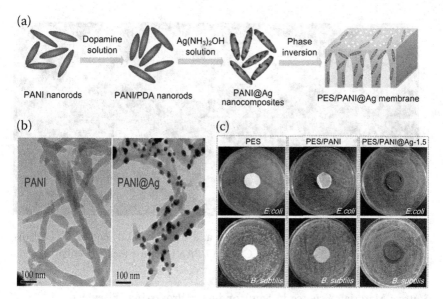

FIGURE 6.8 (a) The synthesis process of PANI@Ag nanocomposites and blending modification of PES UF membrane; (b) morphological characterization of PANI and PANI@Ag nanocomposites observed from TEM images; (c) the inhibition ring results of PES, PES/PANI, and PES/PANI@Ag-1.5 membranes for *E. coli* and *B. subtilis*.

**Source: Biosensors & Bioelectronics (Elsevier).**

numerous cellular components such as lipids and nucleic acids. In addition, Ag NPs can adhere to the bacterial cells and penetrate them, leading to osmotic collapse and release of intracellular materials.

Zhao et al.[38] synthesized PANI@Ag NPs, immobilizing Ag NPs onto polydopamine-coated PANI nanofibers to improve their anti-bacterial ability and stability. The obtained PANI@Ag nanocomposites were incorporated into polyethersulfone (PES) to fabricate composite ultra-filtration (UF) membranes by a phase inversion process (as depicted in the scheme in Figure 6.8a). AgNPs with diameters of ~30 nm were homo-geneously attached onto the surface of PANI nanofibers without agglomeration in Figure 6.8b. The composite membrane revealed excellent anti-bacterial activity by efficiently inhibiting the bacterial growth on the membrane surface. The anti-bacterial activities of PES, PES/PANI, and PES/PANI@Ag-1.5 membranes were evaluated against *E. coli* and *B. subtilis*, and the results of the inhibition ring test are shown in Figure 6.8c, and an improvement in anti-bacterial properties was observed with apparent inhibition zones for both microorganisms.

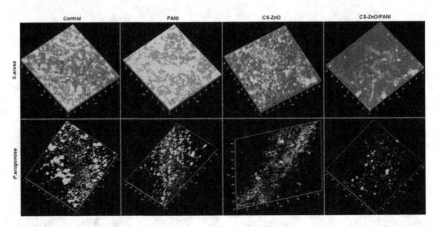

FIGURE 6.9    CLSM images of *S. aureus* and *P. aeruginosa* mature biofilm disrupted by PANI, CS–ZnO, and CS–ZnO/PANI composites on cover glass at their BIC 45 μg/mL.

**Source: Materials Science in Semiconductor Processing (Elsevier).**

K. Pandiselvi et al.[39] fabricated a chitosan–zinc oxide/polyaniline (CS–ZnO/PANI) composite prepared via precipitation with a polymerization method for biomedical applications. Anti-bacterial activities of chitosan–ZnO (CS–ZnO), polyaniline (PANI), and CS–ZnO/PANI composites were determined against gram-positive bacterium, *Staphylococcus aureus* (*S. aureus*), gram-negative bacterium, and *Pseudomonas aeruginosa* (*P. aeruginosa*). Confocal laser scanning microscopy (CLSM) results in Figure 6.9 showed that the control biofilms showed a higher surface coverage while CS–ZnO/PANI-coated surface inhibited *S. aureus* and *P. aeruginosa* biofilm formation by more than 97% and 95%, respectively. Results showed that a CS–ZnO/PANI composite had broad-spectrum anti-bacterial activity that was greatly enhanced in comparison with PANI or CS–ZnO (Figure 6.9).

It would be beneficial in stable and anti-bacterial properties by introducing NPs into the CP's matrix to form the organic/inorganic hybrid composite. The photocatalytic and electrocatalytic activities of the composited materials would be activated or enhanced through energy band hybridization between the conjugated CP molecular and semiconductor NPs, resulting in the bacterial sterilization and pollutant degradation. Examples of the NPs, such as $TiO_2$[34], $MoS_2$[40], $ZnO$[39,41], $n-C_3N_4$[42,43], and $GO$[44–46], ect., have been intensively reported. It should be noted that the photocathode anti-fouling based on a semiconductor and CPs will be introduced in the following (refer to section 6.5.5).

## 6.5.2 Anti-Fouling Properties of PANI-Contained Coatings

Conductive polymers like PANI have a special anti-fouling property along with anti-corrosion potential. The advantages of PANI such as easy synthesis, possessing three distinct oxidation states with different colors, and acid/base doping response along with its cost effectiveness give this conductive polymer credit as a suitable anti-fouling agent for practical applications. Further, the environmentally friendly future also provided another advantage for their wide applications in the ocean.

In this section, laboratory experiments demonstrate that the conducting polymer–based anti-fouling paint is environmentally friendly.[47] It is different from commercial copper-based anti-fouling paint, which leaches out chemicals that are toxic to algae and marine microorganisms. The experiments described by Sze C. Yang[47] showed that although the PANI paint inhibits the growth of algae directly on its surface, it does not stop the growth of algae at the nearby surfaces. In contrast, the commercial copper-based anti-fouling paint not only inhibited the growth of algae on the copper paint surface, but it also killed the algae in the surrounding seawater, and the surrounding non-coated surfaces.

The toxicity of the PANI nanocomposites that contained paint was also performed versus $Cu_2O$ by Mostafaei et al.[48] The epoxy paint containing PANI NPs revealed no evidence of toxicity against algae, while obvious toxicity of copper-based anti-fouling paint was observed. This could be considered a significant advantage for application of PANI-contained marine coating, especially under the background of environmental protection.

The pigments composed of electronic PANI were used to blend into commercial polyurethane and epoxy paints.[47] A laboratory test was performed to verify that the PANI coatings proceed anti-fouling properly. The 5% conducting polymer paints showed resistance to biofouling relative to that of the control sample in a time window of one month before both samples were fouled with algae growth. However, field tests performed in Narragansett Bay in Rhode Island[47] showed that there was a moderate delay of fouling of three weeks to one month compared with control panels without the CP additive. Regretfully, although the CP is a non-toxic anti-foulants, the anti-fouling resistance faded faster than the control panels containing copper paints.

The failure of PANI anti-fouling coating was also noticed after three-month immersion in a simulated conditions similar to marine environment by Fazli-Shokouhi[49], when he investigated the anti-corrosion and anti-fouling properties of epoxy coatings reinforced with PANI/ p-phenylenediamine-functionalized graphene oxide (PGO) composites.

### 6.5.3 Anti-Fouling Properties of the PANI+NPs Contained Coatings

From the above failed cases, it is concluded that although PANI has excellent anti-bacterial properties, as the separate anti-fouling agents used in marine anti-fouling coatings, the anti-fouling efficacy of the coatings still need to be improved. One major disadvantage of the CPs is that the electronic conductivity of the coating slowly decreases due to the dissolution of an anionic dopant in the water. As a consequence of this phenomenon, conductive coating converts into a nonconductive form; thereby, anti-fouling efficiency of the coating decreases. In order to maintain consistent anti-fouling performance, scientists have incorporated various nanomaterials into the epoxy/conductive polymer blends. Various nanomaterials such as $CuO$, $Cu_2O$, $ZnO$, $CeO_2$, $TiO_2$, and GO are operated via different mechanisms to impart long-lasting anti-fouling ability to the coatings.

Ashraf et al.[50] combined the PANI and nano-copper oxide to prohibit the biofouling of polyethylene cage nettings in fish aquaculture. PANI was synthesized in-situ over polyethylene cage netting material and subsequently treated with nano copper oxide. The modified netting material exposed to estuarine environment exhibited excellent fouling resistance. Exposure of the treated cage nets in estuarine water expected to provide vital information on the efficiency of inhibition of foulers in the actual situation. The PE, PE-PANI, and PE-PANI+CuO treated nettings were exposed to the estuarine waters of Cochin estuary for three months (Figure 6.10). The PE-PANI coated with 0.02% copper oxide showed excellent fouling resistance. This study showed that the PANI-coated PE acted as a platform to incorporate nano copper oxide and together these inhibited the accumulation of fouling organisms. Nano copper oxide treated in the matrix acted as the point source above the electron clouds of polyaniline, preventing initialization of biofilm formation and thereby fouling.

The surface of the netting had more active electron clouds of PANI and nano copper oxide that acted synergistically against the accumulation of microorganisms. Further, hypochlorous acid produced by PANI over the

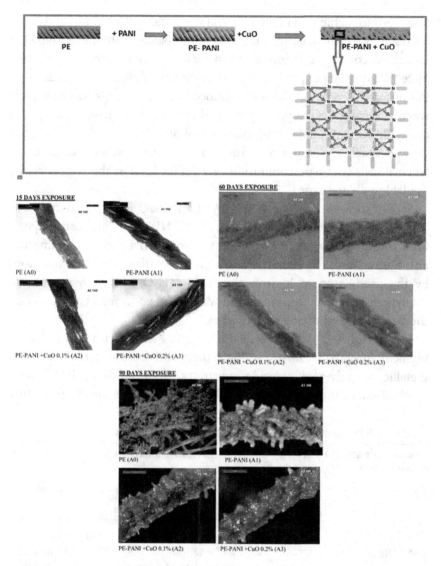

FIGURE 6.10 Microscopic images (25×) of treated and untreated nettings after exposing in the estuary forb15, 60, and 90 days.

Source: Arabian Journal of Chemistry (Elsevier).

surface and biocide activity of nano copper oxide synergistically acted against the attack of microbes, thereby rendering fouling resistance.

Mostafaei et al[48] incorporated nanocomposites of PANI/ZnO nanorods into an epoxy coating matrix. Owing to inherent conductivity of PANI and photocatalytic effect of ZnO, the nanocomposites were observed to exhibit synergetic effects on the anti-fouling and anti-bacterial performances. A hull ship was coated with the PANI/ZnO contained coating, and immersed in seawater for nine months to estimate the anti-fouling performance. The growth of algae and barnacles on the surface of epoxy/ PANI/ZnO nanocomposite coatings was much less when compared to the pure epoxy and epoxy/PANI blend coatings. An epoxy/PANI-ZnO nanocomposite containing 2 wt.% ZnO nanorods showed excellent fouling resistance. Similar results were also observed when PANI/ZnO nanocomposites were embedded in the commercially available polyurethane.[33]

Synergistic effects between the PANI and ZnO were proposed to explain the anti-bacterial properties of PANI/ZnO composited polyurethane or epoxy coatings. While the conductivity of PANI imposes anti-fouling and anti-bacterial effect on the coating, the ZnO functions by the production of hydrogen peroxide ($H_2O_2$) through the photocatalytic reactions with oxygen and water. Thus, formed $H_2O_2$ destructs the bacteria by oxidizing their organs.

Ferreira et al.[22] synthesized PANI and PANI-derivatives as active pigment in epoxy resin to evaluate their anti-fouling performance on metallic or polyvinyl chloride substrates. Four kinds of PANI were synthesized, as shown in Table 6.1. Table 6.2 describes the nomenclature

TABLE 6.1   Preparation Methods of PANI or PANI-Derivatives

| Name of the PANI or PANI-derivatives in Ref.[1,22] | Preparation methods in Ref.[22] |
| --- | --- |
| PANI-ES | PANI-ES was formed by synthesis PANI doped with HCl |
| PANI-EB | PANI-EB was obtained after treatment of PANI-ES with a 0.1 mol $L^{-1}$ $NH_4OH$ solution in water |
| PANI/DBSA | Doping of PANi-EB was achieved in a 0.16 mol $L^{-1}$ aqueous solution of a commercial DBSA solution |
| SPAN | To obtain SPAN, 10g of PAni-ES was dispersed in 300 mL of 1.2-dichloroethane (DCE). $HSO_3Cl$ was diluted to 1.5 M with DCE, and dripped slowly into the PAni dispersion heated to 80°C. The reaction was stirred for 5 hours under reflux. The chlorosulphonated polyaniline was filtered and dispersed in water, heated to 100°C, and maintained at this temperature for another 4 hours to promote hydrolysis. |

*Source:* MATERIALS RESEARCH-IBERO-AMERICAN JOURNAL OF MATERIALS (Open access).

TABLE 6.2    Paints Nomenclature Obtained in the Ref.[22]

| Paint | Epoxy resin | Pigments |
|---|---|---|
| T1-EB | Bicomponent | $Cu_2O$; $TiO_2$ |
| T1-EBPy | Bicomponent | $Cu_2O$; $TiO_2$; $PyZn^*$ |
| T1-EM | Monocomponent | $Cu_2O$; $TiO_2$ |
| T1-EMPy | Monocomponent | $Cu_2O$; $TiO_2$; $PyZn^*$ |
| T2-EMPAniEB | Monocomponent | $Cu_2O$; $TiO_2$; PAni-EB |
| T2-EMPAniEBPy | Monocomponent | $Cu_2O$; $TiO_2$; PAni-EB; $PyZn^*$ |
| T3-EBPAniES | Bicomponent | $Cu_2O$; $TiO_2$; PAni-ES |
| T3-EMPAniES | Monocomponent | $Cu_2O$; $TiO_2$; PAni-ES |
| T3-EMPAniESPy | Monocomponent | $Cu_2O$; $TiO_2$; PAni-ES; $PyZn^*$ |
| T4-EMPAniDBSA | Monocomponent | $Cu_2O$; $TiO_2$; PAni/DBSA |
| T4-EMPAniDBSAPy | Monocomponent | $Cu_2O$; $TiO_2$; PAni/DBSA; $PyZn^*$ |
| T5-EMSPAN | Monocomponent | $Cu_2O$; $TiO_2$; SPAN |
| T5-EMSPANPy | Monocomponent | $Cu_2O$; $TiO_2$; SPAN; $PyZn^*$ |

*Source:* MATERIALS RESEARCH-IBERO-AMERICAN JOURNAL OF MATERIALS (Open access).

of the 13 paints prepared that differ in the nature of epoxy resin used, as well as the amount and type of pigments used.

Anti-fouling tests performed in a marine environment (Figure. 6.11), both in Brazil and in France, showed that with the addition of conductive form of PANI (PAni-ES, PAbi/DBSA, or SPAN) in the paint containing only $Cu_2O$, the anti-fouling performance improved, when compared with coatings containing only $Cu_2O$. The combination of CPs with co-biocide PyZn produces coatings as good as, or even better than, coatings obtained with commercial paints.

And it was concluded that doped polyaniline (PAni-ES and PAni/DBSA) and its sulphonated form (SPAN) can be used as additives in anti-fouling paints, improving the efficiency of anti-fouling coatings, reducing the amount of $Cu_2O$ used in the formulations and hence its release to marine environment. Similar results were described by Wang et al.[21] According to the authors, coatings formed from PAni-ES, with a bicomponent epoxy resin and PAni/DBSA with polyurethane resin, showed a decrease in electrical conductivity after eight weeks of immersion in the sea and also in anti-fouling efficiency. The phenomenon of dedoping of the conductive PANI in the marine environment responded to degradation of anti-fouling efficiency.

Brominated organics have also been reported to improve the anti-fouling properties of PANI-contained coatings. Bisphenol-A epoxy (EP)

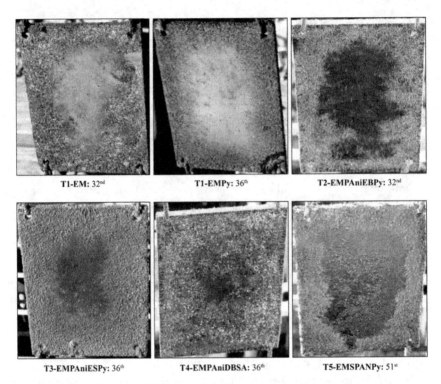

T1-EM: 32nd   T1-EMPy: 36th   T2-EMPAniEBPy: 32nd

T3-EMPAniESPy: 36th   T4-EMPAniDBSA: 36th   T5-EMSPANPy: 51st

FIGURE 6.11   Appearance of coated steel plates and their time of immersion in the marine environment (Toulon - France).

**Source:** MATERIALS RESEARCH-IBERO-AMERICAN JOURNAL OF MATERIALS (Open access).

coating is commonly used to protect metals as a result of excellent scratch hardness, good adhesion properties, etc. However, EP coatings do not possess anti-fouling performance, and may lose their anticorrosion properties under harsh environments in a short time. Hence, bactericides and fungicides are often used in coatings to against bacterial infection [33N]. Tetrabromobisphenol-A epoxy (TBEP) is a bisphenol-A epoxy containing bromine substituents on the benzene rings, and the structure is shown in Figure 6.12b.[51] It had been used as flame retardant resins for a long time, while it has not been paid attention in anti-fouling. Based on the hydrophobic and anti-microbial properties of bromine atoms grafted on benzene rings, TBEP should have multiple functionalities such as anti-fouling, anti-adhesion and anti-bacterial activity. Three types of coatings (pure EP coating, denoted by coating-1; composite coating without PANI, denoted by coating-2; composite coating

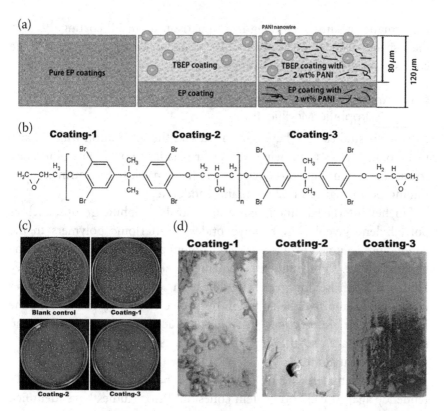

FIGURE 6.12 (a) The compositions of three kind coatings; (b) the molecule structure of TBEP; (c) the fourth anti-bacterial experimental result of coatings against *B. subtilis* after four-week immersion in 12 wt.% NaCl solution; (d) the field test of coatings' anti-fouling performance in river for six weeks.

**Source: Progress in Organic Coatings (Elsevier).**

containing 2 wt% PANI nanowires, denoted by coating-3) were prepared, and the detail as shown in Figure 6.12a. The anti-bacterial test were performed against *B. subtilis*. As shown in Figure 6.12c, apparently an anti-bacterial test demonstrates that TBEP coating has good anti-bacterial properties, while the EP coating is free of it. It needs to be noted that the anti-bacterial properties of PANI in pure PE against *B. subtilis* was not compared in this report. And, the superior anti-bacterial activity of coating-2 and coaing-3 is mainly from bactericidal bromine substituents. The anti-fouling performance of the three coatings were evaluated in the actual environment in the river for six weeks in May and June, as shown in Figure 6.12. The PANI+TBEP PE coating has excellent anti-fouling performance against biofouling in comparison to an EP coating. Besides, the research also demonstrated that PANI nanowires greatly improved the corrosion resistance of the coatings.

### 6.5.4 Anti-Fouling of Polyaniline-Based Coatings with Hydrophilic Modification

Numerous studies[33,36] have confirmed that the integration with CP and metal oxide or NPs can help coatings or plastics to overcome biofouling irresistance in the aqueous environment. Nevertheless, the intrinsic, anti-fouling performance of CPs remains moderate.

Further interfacial functionalization with hydrophilic groups, such as polyethylene glycol (PEG), polypeptides, zwitterionic polymers to the conducing polymer provides several potential strategies to extend the application of the conducting polymer-based films or coatings (see Figure 6.13). The interfacial modification with hydrophilic polymer chains could improve the adhesion resistance of cells by minimizing the intermolecular interactions between extracellular biomolecules and synthetic surfaces so that an adhered cell is readily released under a moderate shear force.

PEGylated materials have been applied because of their strong AF tendency against cell and protein cohesion. Prior studies have identified

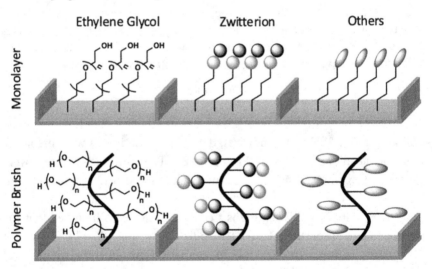

FIGURE 6.13 Design and classification of conducting polymers with an anti-fouling property.

Source: ACS Applied Materials & Interfaces (American Chemical Society).

polyethylene glycol (PEG) as a nontoxic, anti-fouling candidate, and as the sufficiently low interfacial energy (5 mJ/m$^2$) between PEGylated surfaces and water gives rise to the release of proteins and cells.[5,52] Maximizing the surface hydrophilicity and minimizing the attraction forces (caused by formation of hydrogen bonds with water) with fouling community are the mechanistic key issues of PEG.[52]

Zhang et al.[53] describes surface functionalization of the PPy and PANI with poly(ethylene glycol) (PEG) to inhibit the fouling of diatoms, which are a major component of microbial biofouling layers that develop on man-made surfaces placed in aquatic environments, resulting in significant economic and environmental impacts. The capacity of the ICP-PEG materials to prevent settlement and colonization of the fouling diatom *Amphora coffeaeformis* (Cleve) was assayed. Variations were demonstrated in the dopants used during ICP polymerization, along with the PEG molecular weight, and the ICP-PEG reaction conditions, all playing a role in guiding the eventual fouling-resistant properties of the materials. An optimized ICP-PEG material resulted >98% reduction in diatom adhesion.

Among the various hydrophilic groups or hydrophilic chain polymers and zwitterionic polymers, including polyampholytes and polybetaines, have been widely explored as a new generation of fouling-resistant materials. These polymers comprise positive and negative charges, which produce more potent and stabilized ionic bonds with water molecules than those created from other hydrophilic materials.[54,55]

These zwitterionic polymers are promising fouling-resistant materials because of their excellent hydration capacity with strong hydrophilicity. Super hydrophilic pseudo zwitterionic hydrogel, N-isopropylacrylamide (NIPA) + [2-(methacryloyloxy) ethyl] trimethylammonium (TMA) + 3-sulfopropyl methacrylate (SA) co-polymer, is considered a potential anti-fouling agent. Ahana Mohan et al.[56] synthesized nano SiO$_2$-reinforced NIPA-TMA-SA mixed-charged zwitterionic hydrogel over PANI-coated polyethylene (PE) aquaculture cage nets (named as PE-PANI PZH+nanoSiO$_2$) through an in-situ microwave reaction and to test its biofouling resistance. Here, PZH + SiO$_2$ is referred to as a nano SiO$_2$ mixed charged zwitterionic hydrogel. As shown in Figure 6.14, the study highlighted the formation of stable coating over polyethylene, four different treatments, and their effective inhibition of fouling compared to untreated one (refer to Table 6.3). Six months of immersions of treated nettings in the estuarine environments demonstrated that the

One month   Two month   Three month   Six month

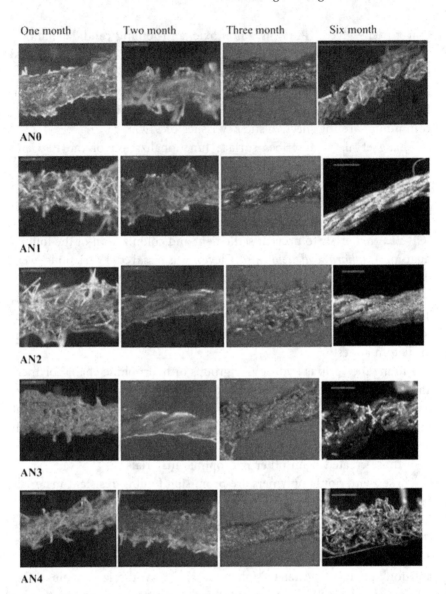

FIGURE 6.14 Microscopic images of the aquaculture cage nets exposed 1, 2, 3, and 6 months in the estuary.

**Source: Langmuir (American Chemical Society).**

biofouling inhibition by nano silicon oxide-reinforced zwitterionic hydrogel coated polyethylene were unable to satisfy the industrial standards but free from hard shelled fouling organisms compared to untreated controls.

TABLE 6.3    Treatment Details of Nano SiO$_2$-Incorporated PZH in Ref. [56]

| Symbol | Treatment | Conc.of SiO$_2$ (%) |
|--------|-----------|---------------------|
| AN0 | PE | 0 |
| AN1 | PE-PANI | 0 |
| AN2 | PE-PANI PZH + nano SiO$_2$ | 0.01 |
| AN3 | PE-PANI PZH + nano SiO$_2$ | 0.02 |
| AN4 | PE-PANI PZH | 0 |

*Source:* American Chemical Society.

To immobilize these anti-fouling groups on conducting polymer surfaces, some groups started with a molecular design for synthesizing monomers with anti-fouling moieties. Anti-fouling conducting polymers can be directly formed using electropolymerization or oxidation polymerization.[38,57,58] Researchers have used surface modification to create anti-fouling layers on conducting polymer surfaces. The anti-fouling layer can be introduced through direct covalent bonding or surface-initiated polymerization methods, such as atom transfer radical polymerization (ATRP) and reversible addition fragmentation chain transfer (RAFT) polymerization.[59–63]

In short, the modification of CPs by a zwitterionic polymer has great potential applications in anti-fouling applications; however, the current related research is mainly focused on the modification of polythiophene and its derivatives, due to the excellent candidate in biomedical related area. Thus, it remains a challenge to surficially decorate the PNAI- and PPy- or their derivative-based coatings with zwitterionic polymer in the marine anti-fouling applications.

Further, in the previous surface modification strategies, both surface-initiated free radical polymerization and electrochemical free radical polymerization targeting conductive polymer monomers involve multiple reaction steps and multiple strong organic solvents, which constrains its application for marine anti-fouling. For example, the involved strong organic solvents will inevitably lead to the dissolution and swelling of coatings, and then damage their anti-corrosion performance. Therefore, from the perspective of practical application, more convenient strategies are urgently needed for zwitterionic surface modification.

## 6.5.5 Cathodic Anti-Biofouling Properties of Conducting Polymers

Cathodic polarization anti-fouling has attracted much attention because of its environmental friendliness. Based on recent reports, it is concluded that conductive polymers represented by PANI and PPy show good anti-bacterial properties, which are directly related to a large amount of $N^+$, redox characteristics, conductivity, and electrochemical activity in their molecular structure. For example, when the CP is in contact with bacteria, a certain cathode voltage is applied to the coating, which may induce an oxygen reduction reaction (ORR) in a solution to produce hydrogen peroxide ($H_2O_2$), hydroxyl radical ($\cdot OH$), and other oxygen active substances, and cause the bacteria to be destroyed by oxidation.

Wang Jia[64] synthesized PANI NPs from a DBSA solution, and investigated the PANI in PMMA coatings as a cathodic material for cathodic polarization anti-fouling in *E. coli*. A remarkable long-time anti-fouling effects of the PANI-PMMA coating under –0.4 and –0.6 V (versus saturated calomel electrode (SCE)) was observed, which was attributed to the production of $H_2O_2$. The excellent cathodic polarization anti-fouling of PPy/acrylic resin coating were also investigated by Wang et al.[65] An Ag dopped PPy (Ag@PPy) composite was synthesized for evaluating the cathodic polarization anti-fouling property Ag@PPy/acrylic resin.[66] Compared with PPy/resin coating, its electrical conductivity was improved by six times due to the Ag decaration. The activity of oxygen reduction reaction (ORR) was also significantly increased, even at a smaller cathode voltage (–0.3 V). Figure 6.15(a–e) show the fluorescence microscopy images of *E. coli* cells on coated samples after polarization at varied potential. As shown in Figure 6.15(a), the PPy/resin sample performed little trace of attached bacteria on the surface after polarization (Ec≥–0.4 V). With the addition of Ag, the concentration of *E. coli* on substrates decreases, obviously, due to the anti-bacterial property of Ag particles. However, the attached *E. coli* dropped precipitously on Ag@PPy-III and Ag@PPy-IV under –0.3 V, quite different from others at this voltage. This may be attributed to the positive shift of reaction (III). Therefore, Ag@PPy-III can achieve the best anti-fouling performance with the least addition of Ag.

Recently, Yu et al.[67] found that dodecyl benzenesulfonic acid doped PPy (PPy-DBSA) could effectively inhibit the attachment of *E. coli* under alternating polarization of cyclic voltammetry anode -cathode (–0.6~0.8 V versus SCE).

From the above description, a conclusion we can draw is that the cathodic anti-fouling is high related to the conductivity and redox activity. Electrochemical redox property of CPs helps in the movement of dopant anti-bacterial ions in and out of the polymer matrix. The unique redox properties of PANI result in controlled ionic transport through the

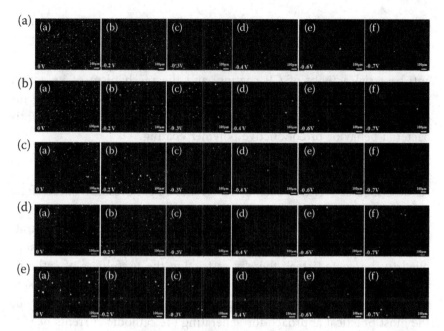

FIGURE 6.15   Fluorescence microscopy images of *E. coli* cells on PPy/resin (a), Ag@PPy/resin-I (b), Ag@PPy/resin-II (c), Ag@PPy/resin-III (d), and Ag@PPy/resin-IV, (e) coated samples after application of (a) 0 V (b) −0.2 V (c) −0.3 V (d) −0.4 V (e) −0.6 V (f) −0.7 for 20 min; potential E (V) versus SCE.

Source: Materials Letters (Elsevier).

polymer matrix to the surrounded solution leading to enhanced antimicrobial effects. CPs are also promising as stimulus response materials because they can be a reversibly oxidized and reduced process that leads to changes in charge, conductivity, and dopant levels, and mechanical and other physical properties. The stimulus response characteristics of CPs have also been explored to show the characteristics of controlled drug delivery.

It should be noted that during the cathodic polarization for antifouling, the overpotential applied should be carefully selected in order to avoid causing the degradation of anticorrosion performance of coating. Thus, attention should be paid to the relationship between polarization potential and coating stability. Further then, from the perspective to reduce the applied polarization overpotential for generating enough hydrogen peroxide or reactive oxygen species, it would be helpful to embed some selected catalytic nanoparticles into the CP polymer to enhance their electroactivity.

In general, there are relatively few reports on the redox activity and anti-bacterial activity of CP coating, and the mechanism of its action with bacteria is not clear. Therefore, it is necessary to conduct an in-depth discussion on the relationship between redox activity and anti-bacterial behavior of conductive polymer coating.

## 6.5.6 Photocathodic Anti-Fouling Properties of CPs

As described above, CPs such as PANI and PPy, have been successfully used as cathodic polarization materials for steel anti-fouling protection owing to their high conductivity. However, cathodic polarization strategies demand that electricity in the form of a cathodic current be supplied to the coated steel, thereby restricting the usefulness of cathodic polarization as an anti-fouling strategy in marine environments. Using semiconductor coatings and sunlight to produce photo-currents may be the most practical approach for generating the cathodic currents needed to achieve cathodic polarization of marine steels. Over the past decade, various approaches have been used to supply photo-generated cathodic currents to metals, with films based on the semiconductors $TiO_2$ and ZnO being the most widely studied due to their low cost and chemical stability.[34]

From the prospective of photocatalyst mechanism, photoexcitation of semiconductor materials (i.e. $TiO_2$, ZnO and g-$C_3N_4$) in photo-sensitive coatings will generate photoelectron-hole pairs (when the energy of the incident light exceeds that of the semiconductor band gap). The photo-induced currents generated in an electrochemical cell, where the semi-conductor coating is part of the working electrode, are sufficient to remove or inactivate biofilms on the electrode or other conductive substrates. Further, the photogenerated electrons and hole can react with $H_2O$, $OH^-$ or dissolved $O_2$ to generate reactive oxygen species (i.e., $\cdot OH$, $\cdot O_2^-$), which involves advanced oxidation processes (AOPs) and which are capable of inhibiting the growth of bacteria.[68,69]

The PANI-$TiO_2$ coatings were found to offer outstanding photo-cathodic anti-fouling and anti-bacterial activities.[34] When PANI is combined with an $n$-type semiconductor like $TiO_2$ ($n$-type due to oxygen deficiency, i.e., $TiO_{2-x}$) the holes generated in $TiO_2$ by under UV photoexcitation are transferred to PANI ($p$-type), while photoexcited electrons generated in PANI under visible light will be transferred into the conduction band of $n$-type $TiO_2$, thus resulting in very effective charge separation and suppression of electron-hole pair recombination in both semiconducting materials. As shown in the cyclic voltammetry curves of Fig. 6.16(b), the current response was much stronger for the PANI-$TiO_2$ composite coating than the $TiO_2$ coating, indicating that the introduction of PANI greatly improved the conductivity of the coating. In Figure 6.16(c), a more negative potential were obtained under light irradiation. For the $TiO_2$ coating on stainless steel (working electrode), the potential realized was only −0.149 V, while −0.41 V for the PANI-$TiO_2$ coating. The value is much negative than the potential required for $H_2O_2$ generation (−0.33 V) indicating that $H_2O_2$ should be formed when the PANI-$TiO_2$ composite coating is under illumination, thereby offering anti-bacterial protection. Further, the light-induced potential was more

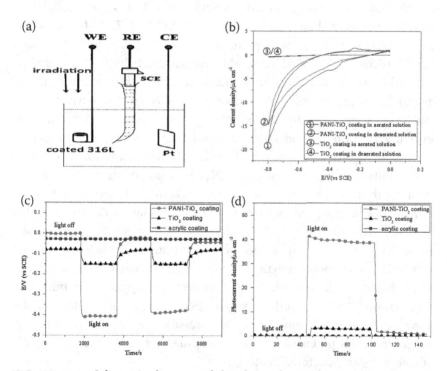

FIGURE 6.16 Schematic diagram of the electrochemical test setup; (b) cyclic voltammograms of $TiO_2$ and PANI-$TiO_2$ composite coatings on stainless steel in aerated and deaerated 3.5 wt.% NaCl solution; (c) potential change curves and (d) photocurrent density curves of PANI-$TiO_2$, $TiO_2$, and acrylic coatings on stainless steel.

Source: Synthetic Metals (Elsevier).

negative than the corrosion potential of 316 L stainless steel (-0.185 V), thus offering a cathodic protection effect (cf. the $TiO_2$ coating which was not capable of offering a cathodic protection). Furthermore, the photocurrent of PANI-$TiO_2$ coating in Figure 6.16(d) was 41.2 $\mu A\ cm^{-2}$ under illumination, which approximately times times higher than that determined for the $TiO_2$ coating (4.71 $\mu A\ cm^{-2}$). The larger photocurrent for the PANI-$TiO_2$ composite coating under light illumination implies that more charge carriers (holes and electrons) were generated in the composite coating than in the $TiO_2$ coating.

Yu et al.[68] synthesized PANI from the surface of silver–N-doped $TiO_2$ composites to fabricate PANI–Ag–N–$TiO_2$ nanocomposites, which were then mixed with an acrylic resin to form coatings on type 316L stainless steel. Under UV–Vis light illumination, the PANI–Ag–N–$TiO_2$ coatings delivered a stable photocathodic potential of –0.79 V, a much larger negative potential than was realized using PANI–N–$TiO_2$ (–0.55 V) or PANI–$TiO_2$ (–0.40 V) coatings. In the system, PANI acted as a visible light sensitizer of N-doped $TiO_2$, while Ag nanoparticles facilitated charge separation across the $TiO_2$/PANI heterojunctions, thus maximizing the availability of photoexcited electrons and holes for reactive oxygen species (ROS) generation and in turn bacterial growth inhibition. Antimicrobial experiments also demonstrated a good anti-bacterial activity (>95% inhibition) against *E. coli* under UV–Vis light irradiation for 30 min.

Graphitic carbon nitrogen (g-$C_3N_4$) is a polymeric semiconductor photocatalyst, possessing a structure consisting of heptazine/triazine units and a bandgap energy around 2.7 eV.[69] Due to its strong visible light response, low cost, high stability, and non-toxicity, g-$C_3N_4$ is now being used as a photocatalyst in many applications, including solar fuel generation and photocatalytic environmental remediation. Recently, g-C3N4-based nanocomposites were successfully applied for microbial inactivation.[42,43,69] Reactive oxygen species (ROS) generated on the surface of the nanocomposites were capable of oxidizing microbes, suggesting great potential for the development of photo-induced anodic anti-fouling/anti-bacterial coatings.

## 6.6 SUMMARY AND OUTLOOKS

It has become the bottleneck to marine resources development, due to the serious corrosion and biofouling of marine engineering installation and transport equipment. Microbiologically influenced corrosion (MIC) and marine biofouling are two main mechanisms of marine corrosions due to the complicated marine environment and marine organisms. To prevent the occurrence of MIC and biofouling, it is important to control the microorganisms in biofilms or prevent the adhesion and formation of biofilms. With the increasing concern to the marine ecological environment, the development of a green, environmental protection, antimicrobial adhesion, excellent anti-fouling agent has become an important research topic in the field of material science.

Conducting polymers and their family of stimuli-actuated polymers have proven to be invaluable not only in the field of anticorrosion but also in anti-biofouling. PANI as a widely studied CP has shown the potential application in anti-bacterial and anti-fouling prospect. In general, the anti-bacterial and anti-fouling performance of polyaniline itself is not very outstanding, and with immersion in seawater, its antibacterial and anti-fouling performance of the PANI coatings would degrade with prolonged immersion in marine water. Some modification strategies can improve its anti-bacterial and anti-fouling performance. The inorganic composites such as Ag, ZnO, CuO, and GO can not only improve coating anti-corrosive properties, but also improve the antibacterial property of the coatings. The PANI used as additives in antifouling paints could improve the efficiency of anti-fouling coatings and reduce the amount of $Cu_2O$ used in the formulations.

The anti-fouling properties of PANI and PPy are directly related to a large amount of $N^+$, redox characteristics, conductivity, and electrochemical activity in their molecular structure. Further, due to their special electroactive property, conducting polymers are also promising as stimulus response materials because they can be reversibly oxidized and reduced process that leads to changes in charge, conductivity, and dopant levels. They have exhibited an application prospect in drug delivery and anti-biofouling. Further investigations to understand the anti-bacterial mechanisms are still required, especially the relationship between redox activity and anti-bacterial behavior of CP coating.

To improve the marine anti-fouling performance, modification of polyaniline is emphasized with hydrophilic chain polymers, such as zwitterionic, and PEGylated materials. However, convenient surface modification strategies for the PNAI- and PPy- or their derivative-based coatings remain still a great challenge.

Further, considering the complex condition in a marine environment, a single strategy is insufficient to achieve anti-fouling effectively. The ideal anti-fouling coatings in marine water must proceed with multiple abilities of anticorrosion, anti-bacterial, anti-fouling and self-cleaning functions.

## REFERENCES

1. Li, Y. & Ning, C. 2019. Latest research progress of marine microbiological corrosion and biofouling, and new approaches of marine anti-corrosion and anti-fouling. *Bioactive Materials*, **4**, 189–195.
2. Heidelberg, J. F., Seshadri, R., Haveman, S. A. *et al.* 2004. The genome sequence of the anaerobic, sulfate-reducing bacterium Desulfovibrio vulgaris Hildenborough. *Nature Biotechnology*, **22**, 554–559.
3. Videla, H. A. & Herrera, L. K. 2005. Microbiologically influenced corrosion: looking to the future. *International Microbiology*, **8**, 169–180.
4. Dong, Z. H., Liu, T. & Liu, H. F. 2011. Influence of EPS isolated from thermophilic sulphate-reducing bacteria on carbon steel corrosion. *Biofouling*, **27**, 487–495.
5. Selim, M. S. *et al.* 2017. Recent progress in marine foul-release polymeric nanocomposite coatings. *Progress in Materials Science*, **87**, 1–32.
6. Lejars, C. B. M. 2014. Marine fouling an overview. *The Journal of Ocean Technology*, **9**, 20–28.
7. Lichter, J. A., Van Vliet, K. J. & Rubner, M. F. 2009. Design of antibacterial surfaces and interfaces: Polyelectrolyte multilayers as a multifunctional platform. *Macromolecules*, **42**, 8573–8586.
8. Little, B. J. *et al.* 2020. Microbially influenced corrosion—Any progress? *Corrosion Science*, **170**, 108641.
9. Chen, L. *et al.* 2021. Biomimetic surface coatings for marine anti-fouling: Natural anti-foulants, synthetic polymers and surface microtopography. *Science of The Total Environment*, **766**, 144469.
10. Xie, Q., Pan, J., Ma, C. & Zhang, G. 2019. Dynamic surface anti-fouling: mechanism and systems. *Soft Matter*, **15**, 1087–1107.
11. Gu, Y. *et al.* 2020. Research strategies to develop environmentally friendly marine antifouling coatings. *Marine Drugs*, **18**, 371.
12. Selim, M. S., El-Safty, S. A., Shenashen, M. A., Higazy, S. A. & Elmarakbi, A. 2020. Progress in biomimetic leverages for marine anti-fouling using nanocomposite coatings. *Journal of Materials Chemistry B*, **8**, 3701–3732.
13. Abdolahi, A., Hamzah, E., Ibrahim, Z. & Hashim, S. 2014. Application of environmentally-friendly coatings toward inhibiting the microbially influenced corrosion (MIC) of steel: A review. *Polymer Reviews*, **54**, 702–745.
14. Raj, J. A., Mathiyarasu, J., Vedhi, C. & Manisankar, P. 2010. Electrochemical synthesis of nanosize polyaniline from aqueous surfactant solutions. *Materials Letters*, **64**, 895–897.
15. Guo, B. & Ma, P. X. 2018. Conducting polymers for tissue engineering. *Biomacromolecules*, **19**, 1764–1782.

16. Yakuphanoglu, F. & Şenkal, B. 2007. Electronic and thermoelectric properties of polyaniline organic semiconductor and electrical characterization of Al/PANI MIS diode. *Journal of Physical Chemistry C*, **111**, 1840–1846.

17. Yang, C. H. *et al.* 2009. Polyaniline/$Fe_3O_4$ nanoparticle composite: synthesis and reaction mechanism. *The Journal of Physical Chemistry B*, **113**, 5052–5058.

18. Genies, E., Hany, P. & Santier, C. 1988. A rechargeable battery of the type polyaniline/propylene carbonate-$LiClO_4$/Li-Al. *Journal of Applied Electrochemistry*, **18**, 751–756.

19. Virji, S., Huang, J., Kaner, R. B. & Weiller, B. H. 2016. Polyaniline nanofiber gas sensors: Examination of response mechanisms. *Nano Letters*, **4**, 491–496.

20. Liu, J., Tian, S. & Knoll, W. 2005. Properties of polyaniline/carbon nanotube multilayer films in neutral solution and their application for stable low-potential detection of reduced beta-nicotinamide adenine dinucleotide. *Langmuir the Acs Journal of Surfaces & Colloids*, **21**, 5596–5599.

21. Wang, X. H. *et al.* 1999. Polyaniline as marine anti-fouling and corrosion-prevention agent. *Synthetic Metals*, **102**, 1377–1380.

22. Baldissera, A. F., Miranda, K. L. d. & Bressy, C. 2015. Using conducting polymers as active agents for marine antifouling paints. *Materials Research*, **18**, 1129–1139.

23. Dhivya, C., Anbu, A. & Radha, N. 2015. Antimicrobial activities of nanostructured polyanilines doped with aromatic nitro compounds. *Arabian Journal of Chemistry*, **12**, 3785–3798.

24. Mu, J. L. *et al.* 2013. The effects of natural dopant acids on morphologies and antibacterial activity of polyaniline. *Advanced Materials Research*, **650**, 249–252.

25. Li, N., Liu, L. & Yang, F. 2014. Highly conductive graphene/PANi-phytic acid modified cathodic filter membrane and its anti-fouling property in EMBR in neutral conditions. *Desalination*, **338**, 10–16.

26. Gallarato, L. A., Mulko, L. E., Dardanelli, M. S., Barbero, C. A., Acevedo, D. F. & Yslas, E. I. 2017. Synergistic effect of polyaniline coverage and surface microstructure on the inhibition of Pseudomonas aeruginosa biofilm formation. *Colloids and Surfaces B: Biointerfaces*, **150**, 1–7.

27. Andriianova, A. N. *et al.* 2021. Antibacterial properties of polyaniline derivatives. *Journal of Applied Polymer Science*, **138**, 51397.

28. Cai, W., Wang, J., Quan, X. & Wang, Z. 2017. Preparation of bromo-substituted polyaniline with excellent anti-bacterial activity. *Journal of Applied Polymer Science*, **135**, 45657.

29. Boeva, Z. A. & Sergeyev, V. G. 2014. Polyaniline: Synthesis, properties, and application. *Polymer Science Series C*, **56**, 144–153.

30. Zang, L., Qiu, J., Yang, C. & Sakai, E. 2016. Preparation and application of conducting polymer/Ag/clay composite nanoparticles formed by in situ UV-induced dispersion polymerization. *Scientific Reports*, **6**, 20470.

31. Zhao, S. *et al.* 2017. Antifouling and anti-bacterial behavior of poly-ethersulfone membrane incorporating polyaniline@silver nanocomposites. *Environmental Science: Water Research & Technology*, **3**, 710–719.

32. Hou, Y., Feng, J., Wang, Y. & Li, L. 2016. Enhanced anti-bacterial activity of Ag-doped ZnO/polyaniline nanocomposites. *Journal of Materials Science: Materials in Electronics*, **27**, 6615–6622.

33. Mooss, V. A., Hamza, F., Zinjarde, S. S. & Athawale, A. A. 2019. Polyurethane films modified with polyaniline-zinc oxide nanocomposites for biofouling mitigation. *Chemical Engineering Journal*, **359**, 1400–1410.

34. Wen, B. *et al.* 2019. The feasibility of polyaniline-TiO$_2$ coatings for photocathodic anti-fouling: anti-bacterial effect. *Synthetic Metals*, **257**, 116175.

35. Kartsonakis, I. A., Liatsi, P., Daniilidis, I. & Kordas, G. 2008. Synthesis, characterization, and anti-bacterial action of hollow ceria nanospheres with/without a conductive polymer coating. *Journal of the American Ceramic Society*, **91**, 372–378.

36. Fazli-Shokouhi, S., Nasirpouri, F. & Khatamian, M. 2019. Polyaniline-modified graphene oxide nanocomposites in epoxy coatings for enhancing the anticorrosion and anti-fouling properties. *Journal of Coatings Technology and Research*, **16**, 983–997.

37. Kim, J. S. *et al.* 2007. Antimicrobial effects of silver nanoparticles. *Nanomedicine: Nanotechnology, Biology and Medicine*, **3**, 95–101.

38. Wang, G. *et al.* 2017. Zwitterionic peptide anchored to conducting polymer PEDOT for the development of anti-fouling and ultrasensitive electrochemical DNA sensor. *Biosensors & Bioelectronics*, **92**, 396–401.

39. Pandiselvi, K. & Thambidurai, S. 2015. Synthesis, characterization, and anti-microbial activity of chitosan–zinc oxide/polyaniline composites. *Materials Science in Semiconductor Processing*, **31**, 573–581.

40. Oturan, M. A. & Aaron, J. J. 2014. Advanced oxidation processes in water/wastewater treatment: principles and applications. a review. *Critical Reviews in Environmental Science and Technology*, **44**, 2577–2641.

41. Krishnaswamy, S., Panigrahi, P., Kumaar, S. S. & Nagarajan, G. S. 2020. Effect of conducting polymer on photoluminescence quenching of green synthesized ZnO thin film and its photocatalytic properties. *Nano-Structures & Nano-Objects*, **22**, 100446.

42. Oves, M. *et al.* 2020. Synthesis and antibacterial aspects of graphitic C$_3$N$_4$@ polyaniline composites. *Coatings*, **10**, 950.

43. Hou, J. *et al.* 2021. Polyaniline/graphite carbon nitride composite coatings with outstanding photo-induced anodic anti-fouling and anti-bacterial properties under visible light. *Progress in Organic Coatings*, **154**, 106203.

44. Fenniche, F. *et al.* 2022. Electrochemical synthesis of reduced graphene oxide-wrapped polyaniline nanorods for improved photocatalytic and antibacterial activities. *Journal of Inorganic and Organometallic Polymers and Materials*, **32**, 1011–1025.

45. Li, B. J. *et al.* 2021. Electrochemical manufacture of graphene oxide/ polyaniline conductive membrane for anti-bacterial application and electrically enhanced water permeability. *Journal of Membrane Science*, **640**, 119844.

46. Kaladevi, G., Wilson, P. & Pandian, K. 2020. Silver nanoparticle-decorated PANI/reduced graphene oxide for sensing of hydrazine in water and inhibition studies on microorganism. *Ionics*, **26**, 3123–3133.

47. Yang, S. C. *et al.* 2009. New anti fouling coatings based on conductive polymers. *University of Rhode Island.*

48. Mostafaei, A. & Nasirpouri, F. 2013. Preparation and characterization of a novel conducting nanocomposite blended with epoxy coating for anti-fouling and anti-bacterial applications. *Journal of Coatings Technology and Research*, **10**, 679–694.

49. Fazli-Shokouhi, S., Nasirpouri, F. & Khatamian, M. 2021. Epoxy-matrix polyaniline/p-phenylenediamine-functionalised graphene oxide coatings with dual anti-corrosion and anti-fouling performance. *RSC Advances*, **11**, 11627–11641.

50. Ashraf, P. M., Sasikala, K., Thomas, S. N. & Edwin, L. 2020. Biofouling resistant polyethylene cage aquaculture nettings: A new approach using polyaniline and nano copper oxide. *Arabian Journal of Chemistry*, **13**, 875–882.

51. Wu, C., Wang, J., Song, S., Wang, Z. & Zhao, S. 2020. Antifouling and anticorrosion performance of the composite coating made of tetrabromobisphenol-A epoxy and polyaniline nanowires. *Progress in Organic Coatings*, **148**, 105888.

52. Kingshott, P., Thissen, H. & Griesser, H. J. 2002. Effects of cloud-point grafting, chain length, and density of PEG layers on competitive adsorption of ocular proteins. *Biomaterials*, **23**, 2043–2056.

53. Zhang, B., Nagle, A. R., Wallace, G. G., Hanks, T. W. & Molino, P. J. 2015. Functionalised inherently conducting polymers as low biofouling materials. *Biofouling*, **31**, 493–502.

54. Blackman, L. D., Gunatillake, P. A., Cass, P. & Locock, K. E. S. 2019. An introduction to zwitterionic polymer behavior and applications in solution and at surfaces. *Chemical Society Reviews*, **48**, 757–770.

55. Acevedo, M. S. *et al.* 2013. Antifouling paints based on marine natural products from Colombian Caribbean. *International Biodeterioration & Biodegradation*, **83**, 97–104.

56. Mohan, A. & Ashraf, P. M. 2019. Biofouling control using nano silicon dioxide reinforced mixed-charged zwitterionic hydrogel in aquaculture cage nets. *Langmuir*, **35**, 4328–4335.

57. Goda, T. & Miyahara, Y. 2019. Electrodeposition of zwitterionic PEDOT films for conducting and antifouling surfaces. *Langmuir*, **35**, 1126–1133.

58. Cao, B. *et al.* 2015. Integrated zwitterionic conjugated poly(carboxybetaine thiophene) as a new biomaterial platform. *Chemical Science*, **6**, 782–788.

59. Wu, J. G., Chen, J. H., Liu, K. T. & Luo, S. C. 2019. Engineering Antifouling Conducting Polymers for Modern Biomedical Applications. *ACS Applied Materials & Interfaces*, **11**, 21294–21307.

60. Wang, J., Wang, D. & Hui, N. 2020. A low fouling electrochemical biosensor based on the zwitterionic polypeptide doped conducting polymer PEDOT for breast cancer marker BRCA1 detection. *Bioelectrochemistry*, **136**, 107595.

61. Zhang, D. *et al.* 2020. Highly stretchable, self-adhesive, biocompatible, conductive hydrogels as fully polymeric strain sensors. *Journal of Materials Chemistry A*, **8**, 20474–20485.

62. Zhang, Y. Q. *et al.* 2020. Tunable protein/cell binding and interaction with neurite outgrowth of low-impedance zwitterionic PEDOTs. *Acs Applied Materials & Interfaces*, **12**, 12362–12372.

63. Zhang, Y.-Q. *et al.* 2021. A trade-off between anti-fouling and the electrochemical stabilities of PEDOTs. *Journal of Materials Chemistry B*, **9**, 2717–2726.

64. Jia, M. Y., Zhang, Z. M., Yu, L. M. & Wang, J. 2017. PANI-PMMA as cathodic electrode material and its application in cathodic polarization anti-fouling. *Electrochemistry Communications*, **84**, 57–60.

65. Jia, M. Y., Zhang, Z. M., Yu, L. M., Wang, J. & Zheng, T. T. 2018. The feasibility and application of PPy in cathodic polarization anti-fouling. *Colloids and Surfaces B: Biointerfaces*, **164**, 247–254.

66. Jia, M. Y., Zhang, J. Y., Zhang, Z. M., Yu, L. M. & Wang, J. 2018. The application of Ag@ PPy composite coating in the cathodic polarization anti-fouling. *Materials Letters*, **230**, 283–288.

67. Zhang, J. Y., Jia, M. Y., Jiang, X. H., Zhang, Z. M., Yu, L. M. & Wang, X. 2019. Antifouling properties of dodecyl benzene sulfonic acid doped polypyrrole under alternating anodic-cathodic polarization. *Chemical Journal Of Chinese Universities-Chinese*, **40**, 2396–2403.

68. Liu, S. B. *et al.* 2020. Enhanced photocathodic anti-fouling/anti-bacterial properties of polyaniline-Ag-N-doped $TiO_2$ coatings. *Journal of Materials Science*, **55**, 16255–16272.

69. Murugesan, P., Moses, J. A. & Anandharamakrishnan, C. 2019. Photocatalytic disinfection efficiency of 2D structure graphitic carbon nitride-based nanocomposites: a review. *Journal of Materials Science*, **54**, 12206–12235.

Printed in the United States
by Baker & Taylor Publisher Services